D1143293

THE
MATING
LIVES
OF BIRDS

THE MATING LIVES OF BIRDS

James Parry

NEW HOLLAND

First published in 2012 by New Holland Publishers (UK) Ltd
London • Cape Town • Sydney • Auckland

www.newhollandpublishers.com

Garfield House, 86–88 Edgware Road, London W2 2EA, UK
Wembley Square First Floor Solan Road Gardens Cape Town
Unit 1, 66 Gibbes Street, Chatswood, New South Wales 2067, Australia
218 Lake Road, Northcote, Auckland, New Zealand

10 9 8 7 6 5 4 3 2 1

Copyright © 2012 in text: James Parry
Copyright © 2012 New Holland Publishers (UK) Ltd

All rights reserved. No part of this publication may be reproduced,
stored in a retrieval system or transmitted, in any form or by any means,
electronic, mechanical, photocopying, recording or otherwise, without
the prior written permission of the publishers and copyright holders.

ISBN 978 1 84773 937 7

Senior Editor Krystyna Mayer
Designer Nicola Liddiard
Production Marion Storz
Publisher Simon Papps
Printer Toppan Leefung Printing Ltd (China)

COVER & PRELIMINARY PAGE IMAGES
Front cover: **Snowy Egret** *Egretta thula* displaying.
Back cover: **King Cormorant** *Phalacrocorax albiventer* breeding pair displaying
courtship behaviour in the form of a ritual dance at the nesting site.
Page 1: **Rainbow Lorikeets** *Trichoglossus haematodus* courtship preening.
Page 2: Male **Blue-footed Booby** *Sula nebouxii* performing courtship 'dance',
blue feet displayed to maximum effect.
Opposite: Desert-dwelling **Gila Woodpeckers** *Melanerpes uropygialis* outside
nest hole in a cactus.

Waltham Forest Libraries	
904 000 00255290	
Askews & Holts	15-Mar-2013
598.1562	£19.99
3821694	

Contents

Adult male **Red Bird-of-paradise** *Paradisaea rubra* preparing to display in a lek tree.

BIRDS' NESTS

How fresh the air the birds how busy now

In every walk if I but peep I find

Nests newly made or finished all and lined

With hair and thistle down and in the bough

Of little awthorn huddled up in green

The leaves still thickening as the spring gets age

The Pinks quite round and snug and closely laid

And linnets of materials loose and rough

And still hedge sparrow moping in the shade

Near the hedge bottom weaves of homely stuff

Dead grass and mosses green an hermitage

For secrecy and shelter rightly made

And beautiful it is to walk beside

The lanes and hedges where their homes abide

John Clare
(1793–1864)

INTRODUCTION

Head held high, bill pointing skywards, a male Song Thrush *Turdus philomelos* pours out his melodious and fluty song from the topmost branch of a mature tree in a British garden. Such is the intensity of the delivery that his throat visibly vibrates as he produces one of the most familiar and recognizable of spring bird sounds in northern and central Europe. To the human sensibility this outburst of singing might seem to be an act of pure joy and celebration, but it is in fact a much more strategic and significant event. It is a statement both of presence, through which the male thrush defines and denotes his territory, informing other males of the same species that this particular area of habitat is already occupied, and of allure, with the singing male advertising to female Song Thrushes that he and his territory are available for mating and the rearing of young.

The act of song is just one component in a much longer sequence of separate but inextricably linked events that make up the avian reproduction process, which if successful ensures the production of adequate numbers of young to maintain a healthy population of each species. Starting with the establishment of a breeding territory and culminating in the successful fledging of independent young, this is a delicate and sensitive process that usually extends over several weeks, if not months or even longer in some cases.

The Song Thrush in full song is a scene reproduced many times over, every day of the year and in almost every corner of the globe, by many millions of birds representing thousands of species. From the pine forests of Siberia and northern Canada to the high mountains of the Andes, through to the African savannah, the lowland rainforest of south-east Asia and the deserts of Australia, the world's 10,000 or so bird species are engaged in a constant drive to reproduce. For many species the use of song is an integral part of that process, but there is a myriad of variations within the vocal arena alone. Birds such as thrushes are well known for being accomplished songsters, to human ears at least, but almost all birds use some form of vocal communication as part of what we might loosely term mating behaviour. In many cases this vocalization can hardly be described as song as such, sometimes comprising little more than a sequence of single notes, for example. Other forms of vocal contact are not made in the conventional manner at all, instead being produced by means such as inflatable air sacs on the neck (prairie-chickens), by the passage of air over and through feathers and wings (snipes) or by the striking of the bill against a foreign surface (woodpeckers).

As pre-breeding activity intensifies, so vocal expressions are increasingly combined with visual communication. This form of activity is equally diverse and often remarkable in nature, encompassing a vast range of different postures, movements and behaviour. These are often – but not exclusively – linked to the manipulation and display of plumage or certain parts of the body, usually those of the males, which in some species develop specific seasonal features to help maximize the effect of the display actions. In many cases such actions follow an elaborate and highly ritualized pattern, which we usually label 'courtship display' and which is an important prelude to the onset of breeding. Particularly well-known examples include dancing (cranes, grebes), arboreal or ground-based acrobatics (birds-of-paradise, manakins), aerial acrobatics (rollers, birds of prey, hummingbirds), lekking, at which

Opposite top Male **Song Thrush** singing. Early spring is the peak time for thrush song. **Opposite below** Grey Crowned-cranes *Balearica regulorum* displaying. Cranes are renowned for their extravagant courtships.

males congregate to perform quasi-territorial displays (grouse, Ruff *Philomachus pugnax*) and even more extravagant activities including the construction – and, in some cases, decoration – of nests and other structures as a way of enticing potential mates (weavers, bowerbirds).

The ultimate purpose of courtship is to secure a mate. Often a seemingly male-oriented affair, it is in reality a process in which the female bird plays a key role for the simple reason that, in most cases, it is the female that selects her mate. Singing and courtship displays are the means by which individual male birds advertise qualities such as health, strength and genetic superiority over other males, all of which are powerful elements in attracting a female to mate. However, the degree of direct participation by female birds in such displays varies greatly between species. In some instances they play as active a role as the males, as in the case of cranes, for example, where both sexes show equal enthusiasm when display-dancing together (see page 66–9). However, the females of most species are noticeably more circumspect. When male Black Grouse

Tetrao tetrix are displaying at a lek, the females usually only frequent the edges of the main arena, quietly observing the action. Yet although they play no direct role in the interaction between males, it is the females who decide on which males to mate with.

During the breeding season the range of behaviour exhibited by birds expands considerably and many acts of courtship display are exaggerated or stylized versions of everyday actions, such as preening or feeding. Normally a bird will preen on its own, but as part of the courtship process and, particularly, the general bonding that serves to underpin the requirement for cooperation between male and female during the breeding season, partners will preen each other (see page 80). The presentation of food offerings is also an integral part of courtship and pair bonding for many species (terns, kingfishers, bee-eaters), yet at other times of the year male and female birds of the same species may compete for food and show little or no inclination to share what can often be a scarce and hard-won resource. The fact that aspects of aggressive and competitive behaviour between the sexes are replaced so

Opposite A female **Black Grouse** bides her time while two males display nearby. She chooses which one to mate with. **Above** A pair of **Great Grey Owls** *Strix nebulosa*. Owls often pair for life and show great site fidelity.

Overleaf Mutual preening, also known as allopreening, is a key part of display and pair bonding, as in the case of these two White Terns *Gygis alba*.

dramatically at the onset of the reproductive cycle shows the potency of the drive to cooperate and breed. Collaboration becomes key, and is prefaced and initiated by the language of courtship with which this book is largely concerned.

What prompts this urge to procreate? Most species of bird breed annually, at particular times of year, with the decision to start that process determined by various interrelated elements. Hormonal changes within a bird are the critical factor in determining the commencement of the reproductive cycle. Several weeks ahead of the breeding season the testes of male birds become greatly enlarged by hormones secreted by the brain in response to factors such as changing day length, rising temperatures and increases in food supply and diversity. Similarly, the reproductive organs within female birds grow in size, eggs begin to form in the ovaries and a brood patch starts to develop (see page 134). These developments are linked to the overall biological condition of a bird, and as it moves into optimal condition, so the urge to breed becomes increasingly powerful.

Most birds split into pairs to breed and stay together for the duration of the breeding season or even longer, but this type of arrangement is by no means universal. Whilst birds such as swans and cranes are monogamous and usually remain with the same partner until one of the couple dies, at the other end of the spectrum are species for which polygamy is the norm and there is no intrinsic recurring bond between breeding individuals. Equally, in the case of most bird species, the female takes the lead in terms of nest construction, incubation and rearing of young, although male partners often assist with the feeding and tending of chicks. However, in some cases males share incubation and in a few exceptional instances it is solely the male's responsibility to rear the young, the female playing no part in the process after she has laid the eggs. Obvious exceptions to the general rule are parasitic birds such as cuckoos and cowbirds — in the case of these birds,

neither parent plays any role at all in the incubation of eggs or rearing of young. This is also true of evolutionary oddities such as megapodes, which rely on alternative sources of heat to incubate their eggs and construct their 'nests' accordingly.

The choices of breeding site and types of nest constructed by birds span as wide a variety of options as the different types of courtship behaviour. Exactly where a bird chooses to lay its eggs, and in or on what, reflects its local environment as much as any other factor, but most birds create some form of nest in which to lay their eggs. The possibilities range from a bare scrape in the ground, with hardly any supplementary material, to the highly complex structures created by weavers, for example. The vast majority of species build an open nest of cup construction, but there are many bird families that have evolved to use other options, from self-excavated cavities to dome-like edifices built of mud or vegetation. A nest needs to provide a secure environment for eggs and also for potentially vulnerable young chicks, yet many birds — especially waders and waterfowl — produce young that are ready to leave the nesting site within a few hours after they hatch. Understandably, these birds spend considerably less energy on creating the nest than do those whose chicks remain within the nest for a number of weeks.

A brood of healthy young birds achieving independence marks the culmination of a season's reproductive cycle. We still have much to learn about the breeding habits of birds, not just in terms of generalities but also with regard to individual species — the reproductive ecology of many birds remains incompletely understood by scientists. Indeed, there is still a surprising number of species for which we have yet to observe the nests and eggs. Yet for many of us, avian mating behaviour is an annual activity that goes on all around us, in gardens and countryside hedgerows and woods. It is a remarkable process, and one that merits both our attention and admiration.

A **Helmet Vanga** *Euryceros prevostii* incubating its eggs. This stage of the reproductive process is highly delicate.

Finding a mate

Securing a mate with which to breed is probably the single most important aspect of a bird's life. The act of mating not only ensures the transmission of an individual bird's genes – something to which some male birds are so tightly programmed that they will destroy the genetic evidence of other males (see page 94) – but also plays an obvious role in helping to ensure the continuation of the species.

The vast majority of bird species are monogamous, in the sense that they will seek to mate and rear young with a single partner in any one breeding season. Although some birds pair for life, most do not maintain such a close relationship, so every year are on the lookout for a suitable new mate with which to breed. Although both sexes play an active role in this process, it is usually the male bird that is responsible for initiating what we may loosely call the courtship stage through the creation of a defined area from which he will seek to entice a female and keep other males at bay.

Summer migrants such as this male **Pied Flycatcher** *Ficedula hypoleuca* usually return to the same area of woodland each spring to breed.

STAKING A CLAIM

Establishing and defending a safe and well-resourced area is a key prerogative for a male bird. This is his offering to a potential mate – a place where successful breeding can take place. Such defended areas, more usually known as breeding territories, vary hugely in size. Small passerines may occupy an area covering less than half a hectare, whereas a territory of as much as 100sq km (39sq miles) is not unusual for the larger raptor species. Some birds, particularly seabirds breeding in what are often exposed and open situations, nest colonially, operating collective territories that afford them easier access to food and better protection from predators. Most male birds defend their territories by patrolling regularly, usually carrying out a circuit of the perimeter, or by singing, calling or displaying from a series of prominent positions where they can be both seen and heard.

Not all territories have the same purpose, however, and the situation is complex and rarely constant in terms of both territory size and function. Many birds maintain feeding territories throughout the year, shifting these as required when the food supplies or habitat within a given area change in some way. Depending on the amount of food available, birds will defend their territories against members of their own kind as well as other species in some cases, and indeed may even seek to prevent their partner from the previous breeding season from utilizing a winter feeding territory. However, some birds – notably birds of prey such as harriers and kestrels – share feeding territories outside the breeding season with others of their species, but zealously chase the same birds out of their breeding territory a few weeks later.

Although resident birds may occupy their breeding area all year round, for migratory species a territory must be established anew each year. In such cases the male birds usually arrive back from their winter quarters earlier than the females and immediately start advertising their presence. By the time the main wave of females arrives, the males have usually arranged themselves into territories. As a general rule, small birds tend to occupy slightly different territories each year, even if they remain in/return to the same general area, but larger (and generally longer-lived) species usually claim the same territory annually. However, the situation depends largely on the level of competition from year to year. In some breeding seasons there may be unoccupied gaps of 'no-man's land' between territories, even in areas of optimum habitat, although new pairs of birds usually quickly occupy vacant spaces, or birds in adjacent territories expand into the gaps if a new male or pair does not take possession. Overall, however, the territory system serves to ensure that populations are evenly dispersed during the breeding season, thereby helping to maximize the chances of each pair successfully rearing young.

MANAGING ASSETS

A satisfactory breeding territory must have certain components. The presence of appropriate habitat – in terms of places for roosting, suitable nest sites for the rearing of young and sufficient refuge from predators – is clearly essential. Adequate access to food and water is important too, although many species regularly feed outside their breeding territories, especially if their food supplies are not easily defendable – as in the case of swallows

Raptors such as this male **Hen Harrier** *Circus cyaneus* regularly patrol their territories to keep out rival males.

and swifts, for example. For many passerines the availability of suitable songposts, from which the male bird defends his territory, is also a key factor in determining territory size and desirability.

Quality of habitat is always variable, with some areas inevitably richer in resources than others. Equally, no two years are the same, so territories often vary in size depending on what is available within their respective boundaries at any particular time. Even within the same species there can be substantial differences in bird density between different locations. For example, suburban gardens can support high densities of breeding birds, encouraged by the provision of food. Today, what we regard as classic garden species like the Common Blackbird *Turdus merula* and European Blue Tit *Cyanistes caeruleus* often occur in greater numbers, and with smaller territories, in gardens than they do in their traditional habitats such as woodland.

For certain groups of bird territory size is determined by special factors. Raptors usually establish relatively large territories, because their food supply is often thinly spread and requires them to hunt over what can be vast areas. At the other end of the spectrum, some territories are remarkably small. In such instances they are often only used for courtship, as is the case with the leks used by the Ruff and gamebirds such as the Black Grouse and prairie-chickens *Tympanuchus* spp., where each displaying male bird defends a very small piece of ground against its peers and uses it as nothing more than a showground at which it seeks to impress females (see page 52). Such lek sites are rarely used for feeding – and they are never used for breeding.

An experienced male bird occupying prime habitat generally stands the strongest chance of mating, as females will assess both his personal attributes in terms of plumage and song (which are signs of his genetic strength) alongside the quality of his territory and the likelihood this combination offers of a successful breeding

attempt. Equally, males of polygamous species like the Western Marsh Harrier *Circus aeruginosus* and Red-winged Blackbird *Agelaius phoeniceus* holding territories with extensive stands of the reedbed cover required by both species for breeding almost always attract more females than males offering more modest options. Male birds living in marginal situations are nearly always less attractive to the opposite sex and often require more extensive territories to ensure that potential breeding requirements are met.

Above Garden feeders can help increase local breeding populations of birds such as these **European Blue Tits**. **Opposite** Noisy and flamboyant, male **Red-winged Blackbirds** are hard to miss as they call and display from reed stems in their marshland habitat.

This entails additional stresses (such as a larger area to patrol and defend, more effort required to find food, etc), which in turn may compromise breeding success even if a mate is found. Males forced to occupy such territories are often immature birds in their first breeding season, unable to oust more mature males from their established territories in the prime sites. The newcomers will await their chance, perhaps when an oldtimer dies, and attempt to occupy his territory when the vacancy arises.

KEEPING CONTROL

Competition for suitable territories can be fierce, and the intensity of territorial rivalry between male birds in spring has long been recognized, with naturalists throughout history recording the animosity and upheaval that the breeding season can bring. The celebrated English naturalist Gilbert White wrote in 1772: 'During the amorous season such a jealousy prevails amongst the male birds that they can hardly bear to be together in the same hedge or field.' Certainly male birds will actively seek to drive out potential rivals from their territory, a process usually achieved by song, visual display or, if that fails, actual pursuit. Border disputes are in fact rather infrequent and usually over very quickly, but in some cases they can result in physical attack.

Certain types of bird have a reputation for being consistently more combative than others. Coots *Fulica* spp., for example, are well known for their noisy territorial spats, which sometimes break out into full-blown fights, often between several males simultaneously. Perhaps most notorious of all are the vicious territorial conflicts that can take place between European Robins *Erithacus rubecula*. Injuries are not uncommon and the most extreme disputes can result in the death of one of the combatants. So zealous can birds be in the defence of their territories that they will seek to drive away not only members of their own species, but also unrelated birds that are superficially similar. For example, avocets *Recurvirostra* spp. actively confront any black-and-white bird that comes near their nest, including oystercatchers *Haemotopus* spp. and the Common Shelduck *Tadorna tadorna*.

Territorial bickering is a daily fact of life for many birds, but nowhere more so than in the case of colonial nesters. Birds like gulls, terns, flamingos and gannets almost always nest collectively, with many thousands of pairs congregating together in what can comprise some of the bird world's most impressive spectacles. The precise level of density can vary, but many tern species, for example, nest in very close proximity to each other, with each pair laying eggs and rearing its chicks just an angry bill's stretch away from its neighbours. In such cases each pair's 'territory' is little bigger than its immediate nest area, but the whole colony can be regarded as maintaining a wider territory that all members will seek to defend against intruders. The second a potential predator such as a skua appears, the colony takes to the wing en masse and attempts to chase it away.

Collective action such as this – which reduces each individual's chance of being predated – is an obvious benefit of colonial living, as are the group feeding strategies often adopted by colonies, whereby the birds leave in flocks in search of food and thereby maximize their individual chances of finding the best feeding sites. But the hurly burly of life in a colony also brings intense pressures to bear. Living cheek-by-jowl heightens aggressive instincts and neighbours are almost constantly engaged in argumentative exchanges, especially so when birds are returning to the colony at the start of a new breeding season. Certain forms of behaviour have evolved to help diffuse such tensions, including various types of ritualized display (see page 63), but even so colonial life is fraught and has a decidedly 'all or nothing' aspect to it.

Opposite above Male **Eurasian Coots** *Fulica atra* are very argumentative and squabbles are commonplace, with birds often using their feet to drive off opponents. Opposite below Seabirds such as these **Elegant Terns** *Thalasseus elegans* typically nest in noisy and densely packed colonies.

THE POWER OF SONG

Once a male bird has identified a potential territory he needs to set about defending it against other males and attracting females with which he can hopefully mate. For birds belonging to the group known popularly as 'songbirds' – actually a distinct minority of species in overall terms – both objectives can be achieved through the means of song. Singing therefore serves two distinct but related purposes, although in the case of most species it appears that the same song in terms of content and delivery is able to achieve both ends. Much remains to be discovered about the subtleties of birdsong, however.

Some birds, such as the European Robin and Northern Mockingbird *Mimus polyglottos*, defend their territory all year round by the use of song. However, most species reserve all or most of their effort for the period immediately before, and during, the breeding season. In temperate latitudes this usually coincides with spring and summer and is an intensely competitive time of year, during which singing is almost always carried out by the male. The sound of a male bird in song triggers the secretion of sex hormones in a female of the same species, which encourage her into breeding condition and prompt her instincts to mate and build a nest. Although the females of some species are indeed known to sing, such activity is usually reserved for the non-breeding season. For most songbird species the female has no song of her own and is simply reactive to that of the male. She assesses it in terms of strength, variety and proficiency of delivery, and makes her choice of mate accordingly. He who sings best usually wins the day.

Not surprisingly, the most enthusiastic and intense singing bouts are generally reserved for the early weeks of the breeding season, when securing a mate is paramount. Many species sing less, or give up entirely, once mating has occurred, although some birds – certain thrushes and finches, for example – may continue singing periodically throughout the summer, and particularly so between broods of young, if the species is multi-brooded.

In tropical areas, where both the pair bond and the territory may be maintained throughout the year and the breeding season is more flexible, singing may be undertaken by both sexes and at any time. However, in almost all situations birdsong has two distinct daily peaks in terms of frequency and volume: a major one immediately before and after sunrise, the celebrated 'dawn chorus', and a second, lesser peak, in the late afternoon and towards dusk. Few birds sing with any gusto during the middle of the day, except when the battle to secure a mate is at its most competitive. At such times birds may sing more or less continuously during daylight hours, losing little opportunity to make their presence felt, although all singing birds leave gaps in their songs so they can listen for replies, whether from potential mating females or from rival males.

Song is affected quite markedly by weather, with warm, sunny and still conditions prompting the most robust singing bouts. Few birds sing anything more than fitfully when it is wet and windy, although there are exceptions. One such is the Mistle Thrush *Turdus viscivorus*, whose tendency to sing vigorously in extreme weather earned it the traditional country sobriquet of 'stormcock'.

Birds vocalize through their syrinx, an organ unique to the bird world that enables them to vary the pitch, tone, rhythm and volume of both their calls and songs. Some birdsong is simplistic and

Top A male **Firecrest** *Regulus ignicapillus* in full song, his brightly coloured crest raised for extra impact. **Above** **European Robins** sing almost all year round, but their repertoire varies according to the season.

repetitive, but at the other end of the spectrum are complicated and very varied arrangements with distinct phrases and an impressive repertoire of notes and sequences. To the human ear, the thrush, lark and warbler families have generally been regarded as featuring the finest songsters, with the Common Nightingale *Luscinia megarhynchos* traditionally accorded the greatest accolades. However, this remains a highly personal choice and every birder will have their own preferred virtuoso.

Although much research remains to be done in this field, it appears that some birds have individually different voices, even dialects or accents. Such levels of individual identity may be important in various respects, not least in terms of territory defence, as recognizing established neighbouring males can save a bird the time and energy involved in rushing to defend a boundary. Where male birds are living in close proximity to one another, regardless of whether territories have already been established, the act of song on the part of one individual often triggers others of the same species to follow suit. Meanwhile, certain species have a very diverse repertoire of phrases and choruses, which some biologists claim is designed to confuse other males into thinking that an area of habitat is already occupied and that they should therefore move on.

Whilst some songbirds combine their song with an aerial display (see page 70), most sing when perched, many from a prominent position where they can both see and be seen. Singing is an acoustic statement related to territory, so it makes sense to ensure that it reaches as wide an audience of rival males and potentially receptive females as possible. However, birds with powerful voices – such as nightingales, wrens and certain warblers – can afford to sing from thick cover. In any case, singing prominently from an exposed location carries risks, including that of being more visible to potential predators.

Above The idiosyncratic 'song' of the male **Corncrake** *Crex crex* has disappeared from much of Western Europe as the species has declined.

The territorial vocalizations of some species do not always accord with the popular perception of what form birdsong might take. Nightjars, pigeons, cuckoos and the enigmatic Corncrake are obvious examples of birds with less melodious songs, yet the distinctive and seemingly monotonous churring of the male European Nightjar – delivered at a rate of up to 40 notes per second – is as much a warning to other males and an enticement to females as the most melodic ditty produced by a thrush. Equally, the bizarre 'song' of the male Eastern Whipbird *Psophodes olivaceus* – a whistle, followed by a whipcrack sound that is magnified by reverberations off the deep undergrowth of the bird's habitat – is not only an integral part of its courtship display but also the first element in a simple duet with its mate, which responds with a two-syllable note. Among the most remarkable songs is that made by the Great Bittern *Botaurus stellaris*, a low-pitched and far-carrying 'booming'. The booms are delivered in a sequence repeated at two-second intervals or so, and in calm conditions can be heard up to 5km (3 miles) away.

NOISES OFF

Birds that do not 'sing' can be far from silent of course, but they often reserve their most vocal moments for actual periods of display rather than indulging in the more extensive broadcast that is typical of songbirds. Male woodcocks *Scolopax* spp. carry out a crepuscular patrol, known as 'roding', over areas of suitable breeding habitat, during which they make various grunting and squeaking sounds designed to attract females. These are only heard during roding. Equally, species that display communally at a lek or other form of arena (see page 52) also utter particular forms of vocalization that are reserved for that activity and place.

Meanwhile some species use non-vocal sounds as part of their territorial and/or courtship behaviour, notably various forms of 'drumming'. Woodpeckers tap or drum on tree trunks and branches in spring to stake their claim to a particular area and to attract a mate. Each species has a particular type of drum, determined by power, duration and the timing between every element, with some birds striking the surface with their bills as much as 20 times per second when drumming.

Another form of drumming is that produced by the Common Snipe *Gallinago gallinago* as part of its aerial display flight. The male snipe flies up rapidly into the sky, calling, then closes its wings and plummets rapidly towards the ground. As it does so, the passage of air through specially adapted outer tail feathers produces a buzzing sound known as winnowing or drumming. In much the same way the Broad-tailed Hummingbird *Selasphorus platycercus* achieves a high-pitched rattle or trill when performing power-dives during its acrobatic courtship flight. The sound is the result of the fast passage of air between an adapted first primary wing feather and its neighbour.

Modified feathers allow certain species of manakin to make mechanical sounds that they use as part of their courtship displays. The Club-winged Manakin *Machaeropterus deliciosus* produces a stridulation noise, rather like that of a cricket, by pulling a curve-tipped secondary feather over a ridged primary at a rate of up to 110 times per second, whilst the White-bearded Manakin *Manacus manacus* has specially evolved wing feathers that make explosive clicks when snapped quickly together. Some birds-of-paradise produce similar effects (see page 60).

A far-carrying non-vocal effect is made by the Ruffed Grouse *Bonasa umbellus*, which beats its wings up to 20 times per second when displaying to produce a muffled drumming sound akin to clapping rapidly with cupped hands. Male Common Pheasants *Phasianus colchicus* perform a limited version of the same technique, whilst Woodpigeons *Columba palumbus* clap their wings noisily together when performing a display flight.

Opposite In early spring European woodlands often resonate with the sound of drumming woodpeckers such as this **Great Spotted** *Dendrocopos major*.

DRESSING TO IMPRESS

With only a minority of birds capable of song, most species have evolved other means of competing in the annual battle to attract a mate. Although many birds conduct their courtship rituals discreetly, and some can be surprisingly inconspicuous during the breeding season, even in these cases the hormonal changes that affect all birds at this time contribute directly to the process of sexual selection.

One of the most noticeable transformations is that undergone by the adult males of many species into what we usually know as breeding plumage, brighter and more colourful than that maintained during the rest of the year. Whilst some species do not have special breeding plumage and look much the same all the time, both sexes need to be in prime condition to help ensure they have the best chance of successfully rearing young.

Plumage changes are achieved by the shedding or moulting of feathers, followed by their replacement with a new set. In the weeks running up to the breeding season many birds undergo a pre-nuptial moult, which may be partial in nature but provides them with fresh feathers that are better able to withstand the strenuous demands of the breeding season. Most bird species usually only moult once a year, but those that adopt a different plumage for the breeding season do so twice.

Pre-nuptial moults result in the males of many species assuming the vivid and extravagant plumage required for effective courtship. This process ranges from the simple replacement of feathers with new ones of a different colour to the growth of special feather features such as crests, ruffs and extended back-, wing- or tail-plumes. Exactly how such refinements have evolved is not clear, but they do play an important role in sexual selection. In the same way that females of certain species choose to mate with those males holding the most desirable territory, so females of birds that use their plumage for display purposes generally select partners with the longest tail feathers, brightest breast feathers, largest crest, and so on. This is again because the extreme development of features such as this is taken as a sign of genetic strength.

In many species the breeding plumage change involves a straightforward upgrading of colour and definition. In spring the male Common Redstart *Phoenicurus phoenicurus*, for example, develops a blacker throat, deeper flame orange underparts and a more clearly contrasting grey and white head. Once breeding is over, he moults into a more muted garb, where the black and orange are distinctly less full and vibrant and the overall appearance is much less crisp. Meanwhile, many birds rely on small areas of coloured plumage, which they may retain all year round, to create impact. Examples include hummingbirds, many species of which have jewel-like iridescent caps, crests or throats, the feathers of which they are able to erect and even 'flash' when aroused or alarmed.

Among the most extraordinary feather adornments are those of the Ruff. In spring the males sport head tufts and extravagant neck feathers, the so-called 'ruffs'. These are very variable in colour, ranging from almost jet black through iridescent dark green and dark brown to cinnamon, cream and white, sometimes flecked or tinted with other colours. Evidence suggests that these colours are inherited by individual male birds from their fathers. Breeding males also develop pronounced coloration on their chests, backs and wings, and shed the feathers on their faces in favour of

Tiny and exquisitely plumaged, the male **Rufous Hummingbird** *Selasphorus rufus* has an iridescent throat patch with which to impress potential mates.

coloured wattles known as caruncles. Outside the breeding season the males assume the same unobtrusive plumage maintained by female Ruffs (known as reeves) through the year.

Other birds also develop crests or feathered adornments to their heads, necks, backs or wings as the breeding season approaches. Many wildfowl species assume particularly spectacular plumage, developing a range of dramatic features, such as the 'sails' of the male Mandarin Duck *Aix galericulata*. He will hold these aloft, and raise his equally impressive crest, when in display. Meanwhile, both males and females of certain grebes undergo a dramatic transformation in late winter, assuming brighter colours and ornate cheek fans, crests and ear tufts in preparation for courtship.

Such extravagant feathers provide the obvious central focus of avian display, but have in the past proved the source of misfortune for the birds concerned, as the human desire to emulate their beauty has often assumed an ugly and macabre dimension. This was never more so than with the vogue in the late 19th and early 20th centuries for bird feathers – or even entire bird skins – as a fashionable addition to ladies' hats. Following the deaths of millions of birds, public outrage at the slaughter eventually put a stop to the butchery and contributed directly to the establishment of conservation organizations such as the Royal Society for the Protection of Birds and the Audubon Society.

The most popular plumes for the millinery trade were those of the Snowy Egret *Egretta thula*, which in spring develops extended, delicately fronded feathers on its back, forming a lace-like cloak, as well as breast plumes and a crest. These are used in both territorial display and courtship, when the male arches his head and neck over his back, fanning the plumes out to maximum effect and engaging in what is known as a 'stretch', in which he points his bill skywards and pumps his body up and down. Both sexes have plumes, with the female responding to the male's lead when in display. Other allies of the heron family, such as spoonbills and certain ibises, have similar features.

Gallinaceous birds such as peacocks and pheasants are particularly celebrated for their ornate appearance, with the males of species such as the Golden Pheasant *Chrysolophus pictus* and Lady Amherst's Pheasant *Chrysolophus amherstiae* exhibiting extravagant neck fans of feathers, which they extend and present sideways-on to potential mates. With crests, iridescent multi-coloured plumage and tails of an outrageous length, these birds are among the most dressy in the entire bird world. The related tragopans of the Himalayan forests even have vivid quasi-luminous lappets with which to dazzle and beguile.

One of the most impressive members of the pheasant family – the Great Argus *Argusianus argus* – is actually one of the less gaudy, relying instead on subtle and cryptic colours and patterns to create a dramatic effect. This remarkable species lives in the rainforests of south-east Asia and has the longest tail of any bird – up to 1.8m (6ft) long, almost twice its body length. Argus feathers are mainly grey and brown, but the elongated and particularly broad secondary flight feathers have a showstopping element: a row of beautifully pigmented oculi or eyespots running alongside each shaft. These have an artful three-dimensional element that works very effectively in the shadows of the forest understorey, and are flanked by swathes of pale and dark spots on contrasting backgrounds. The male makes the most of this spangled patterning in his display, in which he raises his wings by 90 degrees and arches them in a huge circular fan so that his body is obscured behind. The rows of eyespots act like light trails, directing attention to his head, which just pokes out between the wings at the heart of the fan. Meanwhile the two greatly extended tail plumes are raised to appear over the top of the whole ensemble. The sophisticated character of the patterning on the individual feathers must surely rank as one of the miracles of evolution.

Top With both his 'sails' and crest extended, a displaying male **Mandarin Duck** is an impressive sight.

Below left The filigree-like breeding plumes of the **Snowy Egret** were once much in demand by milliners.

Below right Male **Ruffs** sport extravagant neck feathers that they fan out when displaying to each other.

Equally skilled at using feathers to their maximum visual effect are the curious Sunbittern *Eurypga helias* and the world's 26 different species of bustard. The Great Bustard *Otis tarda* is the heaviest European landbird, and each spring mature males undergo a pre-nuptial moult that sees the development of a host of plumage attributes. These include dramatic moustache plumes and elaborate tail and chest feathers, with the latter being brought into play as the bustard's courtship display begins in earnest. The male erects his tail, fans it out and bends it forwards over his back,

drooping his wings almost to the ground and rearing his head back so that his chest is puffed out, to the point at which he appears to have turned himself inside out. The bright white feathers on the chest and wings and under the tail are thereby exposed, and there is even evidence to suggest that a displaying male orientates himself so that his rear end is pointing in the direction of the sun, thereby maximizing the 'gleam' effect – an important consideration when trying to attract females in the bustard's preferred habitat of open steppe.

Above Deep in the rainforest the dramatic courtship display of the male **Great Argus** sees the bird transform itself into a scallop-shaped fan of feathers. **Opposite** A displaying male **Temminck's Tragopan** *Tragopan temminckii* with his bib-like wattle extended.

The spectacular display of the **Great Bustard** takes place in open terrain where it has maximum visual impact.

THE MULTI-MEDIA APPROACH

Birds such as the Great Bustard are largely silent when displaying and rely on plumage and size to impress. The game is moved on considerably, however, by those birds that are not only endowed with magnificent feathers but also bring into play acrobatics and sound effects. Foremost among these are the spectacular birds-of-paradise of New Guinea and northern Australia, which boast some of the most extraordinary ornamental plumage of all. When the first specimens arrived in Europe as preserved skins some scientists refused to believe that any birds could be so colourful or as elaborately plumaged, and suspected a hoax. Almost every group of feathers has been modified into an ornamental variation in at least one of the 40 or so species in the bird-of-paradise family.

Such refinements range from classic lace-like plumes and extended tail feathers through to wiry threads with coin-like discs at the ends, head streamers complete with tiny flags, iridescent panels and even strange enamel-looking sections.

Armed with these often bizarre accoutrements, birds-of-paradise perform remarkable display routines featuring impressive physical gymnastics and a cacophony of outlandish calls (see page 60). Meanwhile, the indigenous peoples of New Guinea have long integrated the birds-of-paradise into their culture and folklore, wearing bird-of-paradise feathers in their head-dresses and even incorporating some of the birds' display movements into their own traditional dances.

Another bird that combines a striking physical asset with a dramatic supporting artillery of movement and sound is the Superb Lyrebird *Menura novaehollandiae*. About the size of a chicken and an otherwise rather ordinary-looking bird, the male lyrebird's claim to plumage fame is a truly monumental tail that is often twice his body length. It comprises 16 feathers known as filamentaries, up to 70cm (28in) long, very thin and filigree-like, and which are framed by two ornate outer feathers ressembling the arc of a bow or a lyre – the so-called 'lyrates'. When displaying, the male fans out his tail and raises it over his body and head like a large umbrella, so that from a distance he almost appears to be in a shower of rain or behind a net curtain. He then hops and jumps about in front of either a competing male or a potential mate in an extraordinary vision reinforced by an accompaniment of seemingly random notes, many of them mimicking other bird species.

More modestly attired birds may rely on a single extreme adaptation to attract the attention of a potential mate, as in the case of two nightjar species, the Standard-winged *Macrodipteryx longipennis* and Pennant-winged *M. vexillarius*. At the onset of the breeding season the male birds of both species develop bizarre extended wing feathers, which trail behind or even above them when they make their nocturnal display flights. In the case of the Standard-winged Nightjar, the feathers – which are the second innermost primaries – can be more than 50cm (20in) in length and comprise a naked shaft for much of this length, with a small section of webbing at each end providing the standards or flags. There are few better examples of such exaggerated feather modification (although see the Marvellous Spatuletail, page 50), but so elongated are these feathers that they rarely survive more than a few weeks, either breaking off or being shed as soon as the breeding season ends.

Above left Male **Magnificent Bird-of-paradise** *Cicinnurus regius* with chest shield extended and tail wires clearly visible. **Top right** The male **King of Saxony Bird-of-paradise** *Pteridophora alberti* has elaborate head streamers up to 50cm (19in) long. **Below right** The bright blue skin on the head of **Wilson's Bird-of-paradise** *Cicinnurus respublica* is clearly visible in the gloom of the forest understorey. **Overleaf** The male **Superb Lyrebird**'s display relies heavily on extravagant tail feathers that take several years to develop fully.

CHANGING PARTS

Breeding season changes in physical appearance can involve more than a set of new feathers. Bill colour and shape can alter, for example, as in the Atlantic Puffin *Fratercula arctica*, the sexes of which look very alike. In early spring, in addition to assuming a crisper and more starkly contrasted plumage, the bills of both male and female are transformed by the growth of brightly coloured external plates. The enhanced bills – the harlequin colour of which may explain one of the bird's traditional names, 'sea parrot' – form a central part of the elaborate pair bonding and display sequences in which puffins engage (see page 86), with the bill plates shed after breeding. Other birds that assume similar bill adornments include the males of duck species such as the King Eider *Somateria spectabilis*, with other waterfowl males developing a fleshy knob-like protruberance at the bill base.

Other features such as facial wattles or lappets, eye-combs, inflatable throat sacs and brightly coloured bare lores (the area between the eyes and bill) are all used by male birds in combination with various forms of locomotion and posturing to impress females. Male prairie-chickens sport bright yellow eye-combs and orange neck pouches, for example, while the male Magnificent Frigatebird *Fregata magnificens* has an unmissable red throat pouch, which he inflates like a balloon and uses as a beacon to attract females.

For sheer physical impact, however, there are few species to beat the Wild Turkey *Meleagris gallopavo*. Almost every part of a mature male in breeding condition is calculated to command attention. The featherless head and neck are covered in fleshy caruncles, with wattles dangling from the neck and the bill partly obscured by a fleshy flap of skin called a snood. When the bird is excited, the bare skin, wattles and snood become engorged with blood and the head undergoes colour changes. Blue signifies sexual arousal, red aggression. The body feathers are puffed up and out, maximizing the appearance of bulk, the tail is fanned out into a semi-circular arrangement and the wings are drooped slightly, creating the impression of an enormous feathered ball.

Some of the most extreme forms of wattle are found on bellbirds, four species of which live in the forests of Central and South America, with all noted for their loud, clanging calls. The male Three-wattled Bellbird *Procnias tricarunculata* has three worm-shaped wattles issuing from the base of his upper mandible; these hang loosely down and are shaken vigorously during courtship. Other forms of wattle are inflatable and serve to amplify courtship calls: the Long-wattled Umbrellabird *Cephalopterus penduliger* has a single feathered wattle of up to 35cm (14in) long and, like other members of its genus, a parasol of feathers on its head that is 'put up' when the bird is aroused.

Exotic breeding garb is most often found in males that do not assist the female much, if at all, in subsequent incubation and care of the chicks. Eyecatchingly colourful plumage would draw unwanted attention to the nest, so males that participate in parental care are often duller and more female-like than males that do not provide help to the female. Even for the brightest males, however, their finery is often only a temporary arrangement. Once the young have fled the nest, metaphorically or otherwise, most adult birds undergo a complete moult, regaining their basic plumage and in most cases – although not all – shedding special

Above In spring **Atlantic Puffins** develop colourful bill plates that play an important role in courtship and pair bonding.

Overleaf **Wild Turkeys** in full display are an imposing sight. Fuelled by testosterone, they can even be aggressive to humans.

45

features such as bill plates. In cases where the sexes look markedly different during the breeding season this period usually sees the male assume an appearance more akin to that of the female, although the situation varies widely. Ducks, for example,

enter what is known as the eclipse stage, a total moult where even the flight feathers are shed, rendering the birds flightless for several weeks. It is all a world away from their pristine appearance of only a few weeks previously.

Above Male prairie-chickens such as this **Lesser** *Tympanuchus pallidicinctus* have colourful neck pouches and eyebrows, which they respectively inflate and raise when displaying.

Above A drake **King Eider** in breeding plumage, showing the bright orange shield above his red bill.

Above The male **Magnificent Frigatebird** attempts to entice passing females by inflating his throat pouch, stretching out his wings and calling raucously.

The art of display

Once equipped with fresh plumage and the basic tools of the courtship trade, most birds are ready to enter the next phase of the breeding process. Songbirds rely on their song to do the business, with some reinforcing its effect with the display of certain parts of their body, either flirting the tail or raising the head feathers or wings.

Particularly extravagant visual displays are mainly the domain of birds with little in the way of a song or coherent call sequence. Among the most impressive are surely the aerial antics of the Marvellous Spatuletail *Loddigesia mirabilis*, a hummingbird found only in a tiny area of Peru. The adult male's tail comprises four specially evolved feathers, two of them bare elongated shafts, curved and culminating in a 'spatule' rather like a flag. When the male is perched these shafts hang beneath him, crossed over, but when he flies up to display to another male or a female, he raises the shafts above his head and whirrs about, flicking the spatules in an eye-catching performance. He is even able to manipulate the two shafts independently.

The male **Marvellous Spatuletail** performs a spectacular aerial display in which he agitates his accentuated tail feathers like tiny flags.

LEKKING

Some of the most fascinating display pageantry is performed by types of grouse, which display collectively in arenas known as leks. This term, which is used to describe the communal display grounds of any species, comes from a Swedish word meaning 'play'. Leks are areas of grassland, heath or open ground beneath trees in which the males of a particular species meet and parade extravagantly in order to dominate one another and court females into mating. These arenas can vary in size from a few square metres to up to a hectare (2.4 acres) or more, and are often used year after year. Male birds come to the lek and display for a few hours on a daily basis during spring, with certain species also attending periodically throughout the rest of the year. There is great site loyalty, with some grouse leks in Scandinavia known to have hosted successive generations of displaying birds almost annually for centuries.

The plains of North America are home to three spectacular lekking species: the Greater Sage Grouse *Centrocercus urophasianus*, and the Greater Prairie-Chicken *Tympanuchus cupido* and Lesser Prairie-Chicken. The hours either side of daybreak are always the best times to watch leks at their most active, with sometimes 30 or more birds in attendance. Leks are species specific, in that birds of different species will not usually occupy the same lek, but in all cases the males hold centre stage. Dominant or alpha males occupy prime position and attempt to fend off the rest, simultaneously showing themselves to best advantage to the females, who gather discreetly outside the arena but with an eye firmly on the action.

Prairie-chicken lekking areas are known as 'booming grounds' and were once widespread across many parts of the United States and Canada. Today prairie-chicken numbers are a fraction of their former size, reduced massively by habitat loss, disturbance and overhunting. When in full display, the elongated neck feathers of the male bird are raised in a ruff around the head, which sinks down in the middle. Particular attention is sought by means of the fleshy ornamental yellow-orange eye-combs and extraordinary neck sacs, which are inflated to full size for maximum impact (see page 48). This process creates the characteristic 'booming', one of the most evocative sounds of dawn on the American prairie.

The Greater Sage Grouse also has an ornamental ruff, which when puffed up resembles a feather boa, and is best known for its characteristic fan-shaped tail of pointed feathers and its inflatable chest sacs, which it inflates and 'claps' together. As is the case with other lekking birds, much of its routine involves prancing, pivoting and strutting around whilst uttering various unlikely but far-carrying sounds. A similar ecological niche is filled in Eurasia by the Black Grouse, which performs various bowing, running and jumping manoeuvures in its attempts to face down other males and woo any interested females. The accompanying bubbling, rasping and hissing calls can carry for some distance across its moorland and heathland habitat.

The pinewoods of northern and central Eurasia are home to the Western Capercaillie *Tetrao urogallus*, one of the largest of all gamebirds. Standing over a metre (3ft) tall, the male bird is an

Above A male **Greater Sage Grouse** in display, his inflatable yellow chest sacs just visible among his chest feathers and with the typical spiked fan tail.

Overleaf Lekking **Black Grouse** males may spend hours displaying to each other, returning to the lek every day for several weeks.

53

imposing sight at any time of year, and especially so when in full breeding regalia. During his courtship routine, usually performed in a clearing in the forest, he fans his impressive tail and points his head upwards, displaying his beard-like throat feathers and flashing dramatic crimson eye-combs. He struts around, occasionally jumping into the air and constantly calling in a series of rasping noises and bizarre 'popping' sounds. Males can chase one another with real aggression, and may show threatening behaviour to human intruders at their leks. In extreme cases testosterone-filled male capercaillies have even attacked vehicles.

When lekking, grouse species often leap off the ground to challenge other males, but some species of bustard perform solo acrobatics to attract the eye of potential mates. The Little Bustard *Tetrax tetrax* in particular has a quirky display that involves puffing up its neck, flicking back its head while giving an unusual 'raspberry-blowing' call, then jumping into the air, sometimes repeatedly, before parachuting down again.

Lekking is not restricted to gamebirds. Two species of wader also use this communal display system: the Great Snipe *Gallinago media* and, most famously the Ruff (see also page 32). Ruff leks are usually located on a raised hummock in open ground, with the first male birds arriving back from their winter quarters in late April – the females usually return a couple of weeks later. A male will stand erect, scanning for a potential opponent. When one is sighted, he rushes forwards to confront it, crouching low with his rear quarters raised and his ruff erected and spread for maximum effect. The other bird will respond in kind, and the two of them may maintain this position, facing each other and motionless, for some time before exploding into a fast and furious 'duel', when they sprint after each other – behaviour that gave the species its German name of *Kampfläufer* or 'fighting runner'. Unusually among lekking birds, Ruffs are completely silent at the lek.

Although Ruffs are highly gregarious and males sometimes gather in large groups at a lek, a displaying male never confronts more than one other male at a time – incidents of males 'ganging up' on an individual are unknown. However, a complex hierarchy exists within the male Ruff community and three types of male have been identified. Research indicates that around 85 per cent of males are typically territorial, attempting to dominate at a lek and thereby attract and mate with as many females as possible. Competition is so intense, however, that the majority usually fails to do so.

Meanwhile, approximately 15 per cent of male Ruffs are termed 'satellites' and do not attempt to establish control at the lek. Instead, they join a territorial male as a 'companion' and display alongside him, hoping to mate opportunistically with females when the chance comes their way and the territorial male is distracted. Evidence suggests that this strategy works well, with satellites fathering as many offspring as territorial males. Interestingly, satellite males almost invariably have white ruffs. More recently the existence of a third type of male Ruff has been identified, birds known as 'faeders', which retain a female-type plumage throughout the year but still manage to mate successfully with females. Their discovery is evidence of how complex sexual selection is in birds and how far we still are from understanding the full picture.

The intensity of a Ruff lek is such that displaying males often appear oblivious to the presence of humans or potential predators. Bird photographer Emma Turner, who was active in England in the early 1900s, remarked that at some Dutch leks the birds could be so 'preoccupied with their own affairs when displaying, fighting, or otherwise amusing themselves [that] … they will allow themselves to be run over by a motor-car'. She also observed that when at the lek, the male Ruff 'is either as motionless as if he were carved in stone, or else he is vibrating like a toy on wires'.

Top left Male **Ruffs** displaying in their characteristic crouching position.
Top right A male **Little Bustard** parachuting down from one of his display
'leaps'. **Above** Two female **Western Capercaillies** eye up a displaying male.

Leks also form a critical element in the breeding cycle of several families of tropical forest-dwelling birds. These include the cotingas and manakins of South America and the celebrated birds-of-paradise of Australasia, all of which may attend their leks throughout the year. The cotingas include many spectacular species, but perhaps none more so than the two species of cock-of-the-rock, the Andean *Rupicola peruvianus* and Guianan *R. rupicola*. As in the case of all lekking species, the males gather in a communal display ground to compete, but in this case the arrangement is rather more formalized into what could be viewed as a form of 'jungle speed-dating'. The males occupy various perches within the lek, each bird keeping to its particular perch most of the time. However, every so often they will attempt to occupy a central position — usually the domain of the dominant male and to which females are most likely to go to mate. Young cocks-of-the-rock start their careers on the outside of the lek, attempting to entice females but acutely aware that it is the central perch that has the most pulling power and to which they must ultimately aspire.

The competitive element that is at the heart of all lekking activity is nowhere more keenly apparent than in manakins, the smallest birds known to use leks. Research indicates that male manakins can spend up to 90 per cent of their day at the lek, and that life there can be very intense. There are almost 50 different species of manakin and the males are often brightly plumaged, albeit without any ornate plumes. What they may lack in such accoutrements is more than compensated for by their antics at the lek. Different species have different displays, and indeed select varying situations in which to perform, with some doing so up in the canopy and others close to, or even on, the forest floor.

Manakin lek sites are carefully selected to meet various requirements, such as opportunities for rays of the sun to highlight vividly coloured parts of the plumage, and the presence of suitable display perches or patches on the ground where individual males can hold court. In favoured spots, 20 or more males may gather, each occupying his own patch (which he tidies up and keeps in good order) and waiting quietly — often for hours — for a female to come within sight or earshot. When this happens, the first male to spot the approaching female starts to display, by perhaps clicking his wings, bobbing up and down, or darting backwards and forwards from one perch to another. His peers follow suit immediately, so that within seconds the entire extended lek is buzzing with sound and movement, a situation that becomes increasingly frenzied as each male advertises himself to the best of his ability. Birds of the same species often have different routines, and individual birds have been recorded rehearsing particular aspects of their display, whether these be short circular flights, little jumps or even a remarkable backwards slide down a branch. However, males do not generally intrude on each other's domains, relying instead on the power of their collective number to attract females to the lek in the hope that they might pass by their own particular court and mate with them.

Related to the crow family, birds-of-paradise are distinctly glamorous compared to their cousins. They are best known for their plumage, but their fabulously ornamental feathers are simply one aspect of their elaborate courtship rituals. The males of most species indulge in display routines that are quite breathtaking in their originality and choreography, with the birds performing actions and making sounds that are unique in the bird world. Many species remain little understood and the nests of some have never been found. Certain birds-of-paradise are now very rare and facing extinction, and the numbers of several others have declined significantly since the Welsh 19th-century naturalist and explorer Alfred Russel Wallace — the first European to see birds-of-paradise alive — saw them in such numbers that he described 'entire trees waving with plumes'.

The bright-red plumage of the **Andean Cock-of-the-rock** helps guarantee visibility in the dark forest undergrowth.

Many birds-of-paradise use leks, but the degree of group courtship varies and some males may even lek on their own. The Raggiana Bird-of-Paradise *Paradisaea raggiana* certainly displays communally, with several males gathering together in a 'lek tree' and performing a vigorous dance with the shimmering plumes that grow from their backs and flanks. These plumes may not develop fully on a male until he is six or seven years old, but young males gather in the lek trees nonetheless and learn and practise their moves in the company of the adults. Shivering, shaking, sometimes rotating himself upside down and giving the impression of a massive and mobile flower in bloom, a male in full display looks as though he is in a trance. At times such is his preoccupation with displaying to his peers that nearby females can end up ignored.

The males usually call loudly to attract the attention of any females in the area before they start their displays, but often move onto a different vocal accompaniment once a female shows some interest. The male Blue Bird-of-Paradise *Paradisaea rudolphi* hangs upside down and pulsates his blue plumes whilst simultaneously rendering a bizarre vibrating sound rather like a muffled soundtrack from a computer game. Equally extraordinary is the cracking sound that the Superb Bird-of-Paradise *Lophorina superba* makes by clicking the quills of his wing feathers together (rather like a human might click their fingers) whilst leaping in front of, and around, a female. Her attention secured, however temporarily, he then dramatically unfurls his black cape (composed of feathers that he raises above his back) over his head and presents his extraordinary iridescent blue-green chest plate to create a bizarre elliptical shape on bouncing legs. Clever footwork is also the hallmark of the male plumed birds-of-paradise, *Parotia* spp. A male will fan out his wings, angle his iridescent chest panel, then sway from side to side, whilst agitating his six head pennants. Even so, success is never guaranteed. Displaying males may try their best, but ultimately it is the females that make the choice.

A male **Raggiana Bird-of-paradise** displaying, his plumes erect and head and shoulders lowered to accentuate the impact.

JOINT DISPLAYS

Most lekking birds are polygamous and will mate with as many females as they can. Their pair bond as such is therefore very brief and the male bird plays little or no part in the rearing of the young (see page 88). However, there are examples of birds that display communally despite being essentially monogamous, either within each season or even for life.

One such group is the albatrosses. These remarkable birds hold a unique place in the human imagination. Whether from Samuel Taylor Coleridge's 1798 *The Rime of the Ancient Mariner*, in which the mariner is famously forced to carry one of these extraordinary birds around his neck, or from more contemporary accounts of the calamitous decline in the populations of many albatross species, these impressive seabirds command our attention.

Wandering Albatrosses *Diomedea exulans* spend most of their lives at sea, only coming to land to breed every other year and only then once they have attained maturity at between 11 and 15 years old. They are monogamous – usually pairing for life – and colonial nesters, generally returning to their breeding sites during the early weeks of the austral summer. With increasing numbers of birds present at a colony, a form of mass stimulation often ensues, with pairs prompted to engage in communal displays. The male and female of a pair face each other with their wings partly spread. Other birds – usually also in their pairs – encircle them, calling while the central pair begins an elaborate sequence of movements. These are characterized by 'sky-pointing', in which both birds point their bills vertically, with bill-clattering and other ritual gestures like the placing of the bill under first one wing, then the other, with each bird usually in synchronicity with its mate. Both birds also march in tandem, in the process often moving out of the central arena and among their surrounding peers. On occasions they even 'retire' and allow another pair to take over. Such communal displays mark only the beginning of the breeding rituals of this species, with a further sequence of related behaviour performed by a pair once it has selected its precise nesting site.

Manoeuvres such as sky-pointing are also an important part of the behaviour of other seabirds, notably gannets and boobies. Northern Gannets *Morus bassanus* nest colonially and usually in very close proximity to each other, so aggression levels are high, especially between males and even between potential mates. Once a male has established himself on a viable nest site, usually on a ledge or a small area on a cliff-top, and almost invariably sandwiched between sites defended by other males, he will stand on it, bowing and shaking his head with his wings half raised. This serves to intimidate other males and make it clear that he is in control of that patch of ground. He then advertises his nest site by displaying while standing on it, and when a female approaches he initially confronts her – as she is essentially invading his territory. The female then turns away and hides her bill as a submissive gesture. Once her subservience is established, the two enter into a courtship ceremony that involves repeated bowing movements and the crossing or 'scissoring' of their bills, actions that further reduce any sense of aggression between the birds. Sky-pointing is also performed as part of courtship, but appears to have even more value as a core element of pair bonding. In the case of Blue-footed Boobies *Sula nebouxii*, the males famously court females by 'dancing', raising their brightly coloured feet as they goose-step their way to and fro, head and bill pointing skywards, tail cocked and whistling all the while (see page 2).

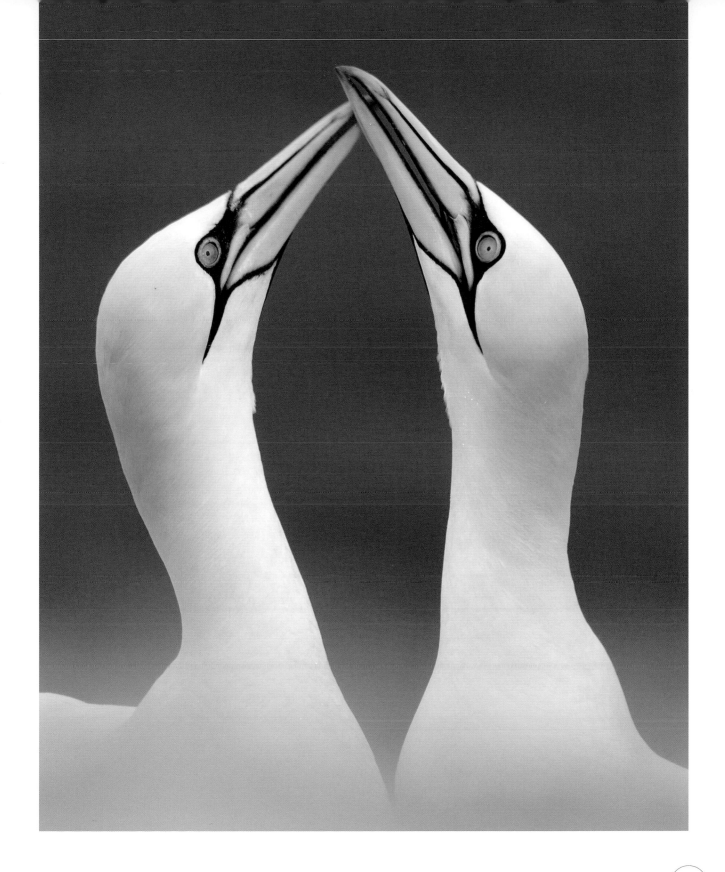

Above Gestures involving the bill are important elements of courtship and bonding behaviour, especially between male and female gannets (here **Northern Gannets**).

Overleaf With a wingspan of up to 3.3m (11ft) **Wandering Albatrosses** need open space in which to display.

DANCING

The performance by birds of what we might anthropomorphize as dancing is not restricted to seabirds. It occurs also in some waterbirds, notably grebes. In the early 20th century the English naturalist Julian Huxley studied the courtship behaviour of Great Crested Grebes *Podiceps cristatus* and identified the various different phases and components. These include a defensive posture called the cat display, adopted by breeding birds when threatened, in which the head is lowered, the elaborate ruff and head-tufts are raised and the wings outstretched. Better known are the mutual head-shaking that pairs undertake together in preparation for breeding, and the birds' celebrated 'penguin dance', in which both birds raise their torsos out of the water to face each other, each holding a piece of waterweed in its bill. Both sexes play equal roles in this ceremony. In North America the Western Grebe *Aechmophorus occidentalis* also performs a version of the penguin dance but has an even more exciting display phase, one in which both male and female run together across the surface of the water in synchronicity with each another, their heads and necks held in a cobra-like posture, before both plunge underwater.

There are also synchronized elements within the collective display activity that is a hallmark of all the world's six species of flamingo. These strictly colonial birds have no regular breeding season, and populations may not breed at all in certain years. Whether they do or not appears to be linked to rainfall levels and the ensuring availability of adequate food, but one of the signs that a breeding attempt is likely are the ritualized group displays that can involve hundreds of birds simultaneously. Performed by both sexes and accompanied by noisy vocalization, these displays begin several months before eggs are laid and comprise sequences of neck-stretching, head-shaking and nodding – usually called 'flagging' – as well as what are known as wing salutes, in which the birds extend their wings and reveal the highly contrasting wing pattern. The performance culminates with the spectacular so-called flamingo dances or marches, when a group of birds marches as one in a particular direction, before turning abruptly and doing the same in another direction. The whole process serves to bring a flock into reproductive condition at the same time, so that most eggs are laid when local conditions appear most favourable for the successful rearing of young.

Dancing of a less regimented nature is also a famous part of the courtship behaviour of cranes. Traditionally regarded as a harbinger of spring and deeply ingrained in the traditional cultures of Eurasia and North America, the spring mating ritual of cranes has a particular resonance. The stately nature and sheer size of these birds, often standing 1.8m (6ft) tall, accompanied by their highly vocal nature, never fails to capture the imagination. There are 15 crane species worldwide, some of which are long-distance migrants, and all of which are known to perform courtship dances and displays.

Although many crane species are under threat from disturbance and habitat destruction, the numbers of the Common Crane *Grus grus* of Eurasia seem to be holding up well and they are even

Opposite top A group of **Lesser Flamingos** *Phoenicopterus minor* 'dancing' in a form of regimented display. **Opposite below** **Western Grebes** perform a spectacular synchronized sprint across the water.

increasing in some places, such as Britain. They require extensive areas of wetland in which to breed and although they are highly gregarious for much of the year, in early spring individual pairs – which usually mate for life – establish a territory and start to defend it against others of their species. Part of this process involves 'dancing', in which birds jump off the ground, trumpeting loudly with their wings half-raised. Extravagant head movements and bill-clapping can also take place. This ritual not only signifies to other cranes that a particular area is occupied, but also helps cement the bond between the male and female.

In late winter and early spring cranes often perform their courtship dances in groups, with whole flocks employing ad hoc routines. This can make for spectacular viewing, but has a serious element: not only does it enable established pairs to reaffirm their relationships, but it also gives younger unmated birds the opportunity to find a partner. The courtship behaviour of the Red-crowned Crane *Grus japonensis* is one of the most elaborate of all: the routine usually starts with individual pairs of birds standing side-by-side and trumpeting loudly in what is known as unison calling. This is followed by a series of bows, head-shaking and bobbing movements before the birds start walking, then prancing about, wings half-raised and heads often turned skywards and even lain over the back. Small leaps into the air become ever more exuberant jumps, accompanied by frenzied wing-flapping and constant calling. In moments of particularly intense arousal props may be used, with birds picking up grass and sticks and tossing them about.

Common Cranes are exuberant dancers, with both pair and group interaction characterized by regular bouts of jumping and calling.

EXPRESSIVE BEHAVIOUR

For cranes, dancing is not restricted to courtship and breeding. It can occur at any time of year and even younger birds of non-breeding age take part. It is clearly part of wider crane behaviour and may help thwart aggression, relieve tension and stress within a flock, or simply be an expression of relief: cranes often dance shortly after arrival at their wintering grounds, for example. In the same way, many bird courtship activities are stylized extensions of everyday behaviour. They involve particular sets of movements that can be traced to the displacement activities that are part of daily avain life, such as head-flagging, preening, bill-wiping and scratching, all of which birds do when they are nervous, unsure of a particular situation or wishing to appease another.

This can be seen clearly in the display behaviour of certain waterfowl, for example. Male ducks frequently preen vigorously as part of their display, using the opportunity to highlight especially ornate aspects of their plumage. In such cases, the preening is primarily symbolic rather than functional. Other forms of duck display have their origin in what are known as intention movements – the preparatory stages for particular actions such as flying, running or calling. For example, when courting females both Common Goldeneye *Bucephala clangula* and Hooded Merganser *Lophodytes cucullatus* drakes throw their heads back in a jerking movement that is essentially a stylized and exaggerated version of part of the take-off leap undertaken by all ducks when launching into flight. Other bird species display by repeatedly nodding their heads up and down.

Although many species of raptor are surprisingly discreet and understated during courtship, others engage in aerial displays as part of their territorial and pre-nuptial rituals. Especially spectacular is the 'sky-dancing' performed particularly by harriers and certain species of buzzard, in which the bird – usually, but not exclusively, the male – takes a long glide, then swoops upwards before stalling, hovering briefly with the wings quivering, then descending equally rapidly before repeating the movement again. In some species this rollercoaster manoeuvre has been recorded as lasting for up to 30 minutes at a time, with up to 40 'peaks' made. Some species also engage in 'talon-grappling', where the male and female lock talons and tumble and somersault through the air together before disengaging. Eagles and harriers in particular are known for their dramatic food passes, in which the male approaches the female with an item of prey in his talons, flies above her, then drops the item for her to catch as she rolls over on her back, legs extended, to catch it.

Some passerines specialize in courtship flights during which their song, or part of it, is delivered in combination with an aerial display of varying degrees of sophistication. Several species of swallow, lark and pipit perform such displays, including the Woodlark *Lullula arborea*, which delivers its soft, fluty song variously from the ground, while perched and on the wing. The Tree Pipit *Anthus trivialis* has a particularly impressive parachuting display flight, climbing steadily upwards before descending quite suddenly with its wings half-raised. Species of wader also undertake regular display flights over their territory, either towering upwards then descending gently, or tumbling through the air while calling repeatedly.

A male **Common Goldeneye** uses his brightly coloured head when displaying, bobbing it up and down or laying it over his back.

SPECULATIVE BUILDING

The ritualizing of behaviour and its incorporation into display additionally extends into nest construction. Penguin courtship involves much bowing and posturing, and is very much part of the wider panoply of penguin etiquette that is perhaps the only way in which colonies of, sometimes, millions of birds can function. Male Adélie Penguins *Pygoscelis adeliae* arrive on their breeding grounds in October each year, usually in advance of the females and having walked many kilometres across the pack ice and rocky terrain. They prospect for nesting sites immediately, gathering together pebbles as the rudiments for the 'nest', which they then stand on or near. When a female approaches to inspect, the male may offer her a pebble as an incentive to choose him. Her acceptance of the gift usually indicates her acquiescence to him as a mate, but there may still be rocky times ahead. Theft and skullduggery are commonplace, with the males often stealing particularly desirable pebbles from their neighbours and the females sometimes proving impossibly fussy in terms of what they will accept. They often reject certain pebbles brought to the nest by the male, tossing them aside, and endless collecting trips may be required by him before the nest is finished and the female satisfied.

Some of the most sophisticated and complex nests of all are built by weavers, with certain species using nest construction as an integral part of their courtship. Males weave a nest, then position themselves on or near it, advertising their presence and that of their finely crafted handiwork by sometimes dangling upside down from the nest, flapping their wings madly and chattering incessantly. They may also periodically add new material and carry out improvements to the nest, as if providing further evidence of their skill and commitment. Because weavers are colonial nesters, there may be very many males doing the same thing, at the same time, in the same tree. A female will come and make her choice, and it is her responsibility to put the finishing touches to the nest by lining the inside. Meanwhile, the male will have mated with her before leaving that particular nest in order to construct another one, in the hope of enticing a new female. Males may make several such nests, and mate with several different females, in any one breeding season.

Some species have taken ritualized nestbuilding as part of courtship one step further, to the point at which each male provides a potential mate with options. A number of wrens do this, for example, with the male seeming to have an irresistible drive to build. He usually constructs several different nests within his territory before attempting to entice a female on a guided tour of his endeavours with a view to her making a choice. Interestingly, usually only one of the nests is properly constructed, the others almost acting as decoys. The female generally chooses the soundest one, an example perhaps of sexual selection cutting both ways – if she settles on one of the decoys, the male may not go ahead with the pairing, as such a poor choice can indicate her lack of experience or unsuitability for rearing a family successfully.

Adélie Penguin males laboriously build speculative nests of pebbles with which to entice prospective mates.

COURTSHIP THEATRES

The speculative construction of nests with which to entice a mate is one thing, but the creation and presentation of a separate arena dedicated solely to the potential act of mating seems an extraordinary investment of time and energy. Yet that is precisely what the bowerbirds of Australasia do. Generally dull in colour, bowerbirds are seemingly at the other end of the bird spectrum from, say, ornately plumaged avians such as birds-of-paradise. In fact, in different ways the two groups represent the same sort of extreme evolutionary response to the intense pressure of sexual selection, with the decorated bowers created by bowerbirds as excessive in their own way as the extravagant plumes of the birds-of-paradise. Early European travellers in Australia did not at first accept that these structures were made by birds, believing them instead to be the creations of Aboriginal people. Bowers are works of art, and the care and originality with which they are decorated is unique in the animal world.

There are 20 bowerbird species, all forest-dwelling, and each creates a different type of bower, of varying degrees of sophistication and decorated with an extraordinary range of objects. The most basic arrangement is that made by the male Tooth-billed Bowerbird *Ailuroedus dentirostris*, which does not build a bower as such, preferring instead to simply clear an area of ground as his 'stage', then decorate it with leaves. Like other bowerbirds, the Tooth-billed male uses a variety of calls to attract females, usually perching above his stage and descending when a female is tempted in. She inspects the bower, decides whether or not the builder is up to the task, then couples with him, or not, before leaving. He mates with as many females as he can during a breeding season that runs roughly from October to March.

Much more elaborate are the structures erected by species such as the Golden Bowerbird *Prionodura newtoniana*, an unusually bright bowerbird in plumage terms. Although it is the smallest species, it constructs the largest bower, a maypole structure made from towers of sticks that can reach up to 3m (15ft) in height and is fitted with a display perch and carefully placed cross-sticks that serve as shelves on which items such as flower petals, seedpods and lichen are assembled. Great care is taken over this process, with the male bird placing an object, then standing back to assess whether or not it 'works'. If he decides it does not, it is moved or replaced with something different.

Certain bowerbirds species show a marked preference for particular objects and even for special colours. Satin Bowerbirds *Ptilonorhynchus violaceus*, for example, favour blue and yellow items. Individuals of the same species may also have varying tastes. The very ambitious bowers, such as those of the Vogelkop Bowerbird *Amblyornis inornatus*, resemble stage sets in their scale and scope, with separate sections reserved for particular objects. These can include drink cans, pieces of plastic, glass shards and even children's toys, collected opportunistically. Bowerbirds may steal prize objects from competing males and display them in their own bowers, although few male birds spend much time away from their stages when there is the chance of a female dropping by. They may even equip their bowers with larders of food, so that they do not have to abandon their posts to go foraging.

Top Male **Satin Bowerbirds** like to adorn their bowers with objects that are blue or yellow, all carefully arranged. **Above** The **Vogelkop Bowerbird** creates one of the most theatrical bowers of all, complete with discrete sections and a proper roof.

COPULATION

The ingenuity and creativity of birds in their courtship endeavours demonstrates how powerful the hormonal drive is for a male bird to find a receptive female and to mate with her. The culmination of bird courtship – the actual act of mating – can often appear brief, perfunctory and even brutal. Copulation takes place with the male mounting the female from behind and with both sexes holding their tails to one side so that their cloacae come into contact. Some birds, such as ducks and gamebirds, have an erectile penis within the male's cloaca, but in most cases the male's sperm is transferred via a short 'cloacal kiss'.

Copulation can take place on the ground, on a perch, in water or even, in the case of swifts for example, in mid-air. Many birds only copulate once with each mate, but some, including birds of prey, may do so more than 100 times over a period of several days. Such differences may be accounted for by the nature of the relationship between male and female and the extent to which they are in each other's constant presence. Raptor pairs, for example, may spend periods away from each other while hunting, so the repeated copulation may be a way for the male of the pair to ensure that his sperm predominates over that of any intruders that may have mated with the female during his absence. Fertilization is not necessarily immediate, however, as the female can 'store' sperm for several days. One other key determinant in the frequency of copulation is the nature of the breeding arrangements of a species. These vary hugely, and cover as wide a range of options as the spectacle of courtship itself.

Above Duck copulation (**Mallards** *Anas platyrhynchos* are shown here) usually takes place on water and can result in the drowning of the female in extreme cases. **Opposite** **Common Kingfishers** *Alcedo atthis* mating in an encounter that may take a couple of seconds before everyday life is resumed.

Personal relationships

Once mating has occurred, preparations for egg-laying begin. However, exactly what happens after copulation depends on the nature of the relationship between the sexes. Roughly 90 per cent of birds are functionally monogamous, a far higher percentage than in other vertebrates, and partly a result of the fact that eggs require continuous incubation and recently hatched young usually need to be brooded and fed. Having both parents helping out is clearly an advantage. However, the partial or total absence of one parent – usually, but not always, the male – is clearly not such an impediment that it prevents successful reproduction. Some species mate promiscuously and the male (or, in some species, the female) plays no part at all in the rearing of young, yet adequate offspring are still reared successfully and the species is sustained without any apparent problems.

Relationships between many male and female birds rely on constant reaffirmation of the pair bond, as in these 'billing' **Tufted Puffins** *Fratercula cirrhata*.

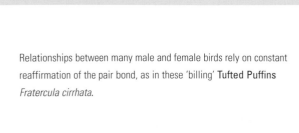

STAYING TOGETHER

Monogamy among birds is not always so much to do with fidelity to a particular individual as to recurring links to a particular place. Whilst some birds – cranes, albatrosses, swans and geese, for example – do pair for life, the vast majority of monogamous species only remain in pairs for one breeding season at a time. However, by residing in the same area for much of the year, or by returning to it each spring in the case of migratory species, the chances of them renewing their bond with a mate from the previous year are reasonably high. Mating with the same partner year after year has obvious advantages, once the arrangement is tried and tested and proved to be successful. Experienced birds are most likely to rear more young, so there is little cause to break a partnership with a good track record. 'Divorce' is unusual, and generally related to recurrent breeding failure.

Even among monogamous species, the precise ways in which parental duties are shared vary widely. In the case of the Garden Warbler *Sylvia borin*, both parents play an almost completely equal role, with the male and female sharing nest-building, incubation and the feeding of young. In other species, including the Blue Jay *Cyanocitta cristata*, the male helps build the nest and feed the young, but does not play a role in incubation. Meanwhile, the males of tanagers, for example, do little after mating except bring food to the female when she is incubating, but then play a full role in feeding the hatchlings. An extreme adaptation of a monogamous arrangement can be found among certain waders and some gallinaceous birds such as the Red-legged Partridge *Alectoris rufa*. During especially bountiful periods, when food is abundant and weather conditions are favourable, the female may form a second clutch of eggs almost immediately after her first. In such cases she then leaves the original clutch for the male to incubate and goes off, makes a new nest and lays a second batch of eggs in that, incubating those herself.

The period immediately after mating is especially important for male birds. Any male will be wasting his energy, time and genetic resources if the female with which he has mated then mates with other males, so the defence of 'his' female is of paramount concern. The males of certain species may steadfastly accompany their females on expeditions to collect nest material (although they will not necessarily assist with gathering it), or even to feed or drink, not out of any sense of affection or loyalty, but simply to ensure that no other male is able to mate with her.

BONDING

Whichever system is followed by monogamous pairs, there is one important element to the arrangement: the need to cement and maintain the parental bond between male and female. In what has been described as an 'uneasy truce' between birds that might otherwise be in conflict over resources such as food, it is important that the consensual agreement to cooperate in rearing young is kept intact throughout. This is achieved by various means, which are either part of everyday behaviour or adaptations of regular actions that can be tailored to suit this particular purpose. They form social signals that appease, reduce or redirect antagonistic tendencies and convert these to bond-strengthening postures.

Foremost among these is social preening, also known as allopreening. Some 40-odd families of bird are known to engage in

Mute Swans *Cygnus olor* usually pair for life and stay together all year round. A partner's death may produce symptoms akin to grief in the surviving bird.

this activity, which takes the form of one bird preening the plumage of another. Such attention is usually restricted to the head and neck, parts of the body that birds find difficult to deal with themselves. The function of allopreening therefore has a clear practical dimension, as it can help birds rid each other of parasites such as lice and ticks. However, mutual preening also has a social function, serving to bond pairs and even groups. Birds will present themselves in special invitation postures, encouraging their mate (or peers) to preen them by ruffling up their own feathers then turning the back of their neck towards the potential preener. Parrots, albatrosses, crows, herons and some finches all regularly do this, and while such behaviour sometimes forms part of the pre-nuptial courtship process, it appears to be more regular as part of ongoing bonding between pairs. Lovebirds are particularly avid allopreeners, seemingly affectionate behaviour that has helped earn them their English name.

Another bonding activity firmly rooted in a practical dimension is gift presentation, which usually takes the form of food items offered by the male to the female. The male's general ability to supply food can be an important consideration to the female when she is assessing his potential suitability as a mate. What is sometimes known as courtship feeding can therefore play a role in the pre-nuptial process. However, in many species it only takes place once the pair has mated and settled down to breed. It can then occur regularly up to and including the period of incubation and can be valuable to the female in the weeks immediately after mating, when the eggs are forming in her body and she needs as many nutrients as possible. Such feeding has an equally obvious practical function in species where the female does most of the incubation or is even incarcerated in her nest-hole, as in hornbills (see page 110), for without the male regularly bringing food she would become undernourished or starve completely. Birds of prey engage in aerial exchanges of prey as part of courtship (see page

70), and ceremonial feeding is especially pronounced in birds such as kestrels, the males of which will regularly present their females with items of food. They usually do this either by bringing the food to a perch near their mate or, if she is incubating, to the nest itself, calling and making a fluttering flight to attract her attention. She often then perches next to the male and the couple may bow several times before the female takes the offered item – a rodent, small bird, grasshopper or beetle, for example – from her partner's bill. This approach is also found in terns, which may preface the exchange of food with a display flight. However, in other seabirds, including gulls, the male simply alights near the female and regurgitates the food for her to take up, in very much the same way

Above A male **European Bee-eater** *Merops apiaster* presents his mate with an insect, behaviour that is part of both courtship and pair bonding.

Top Mutual preening is an integral element of pair behaviour among parrots such as these **Great Green Macaws** *Ara ambigua*. **Above** A female **Wood Warbler** *Phylloscopus sibilatrix* 'display-begging', as she encourages her mate to offer her food items.

Overleaf Pair-bonding displays are regularly observed among **Herring Gulls** *Larus argentatus* – here the Baltic form *omissus* – and involve synchronized movements and action mimicry.

as the pair will later feed their own young. Although it is usually the male that presents the female with food offerings, in some bird groups – notably hummingbirds – these roles are reversed.

Another example of bonding behaviour is 'billing', in which male and female touch their bills together. Frequently described in anthropomorphic terms as birds 'kissing', this activity is almost certainly an advanced ritualization of feeding actions, to the point where it has no discernible practical function in relationship to its origin. Yet nor is it purely symbolic either, for it has assumed a practical element: the cementing of the pair bond. Puffins are especially known for billing, with pairs regularly indulging in this clearly intimate behaviour. It is used not only as part of puffin display, alongside sequences of parading around, head shaking and wing flapping, but also as a greeting and vehicle for pair bonding. Interestingly, a pair engaged in billing often attracts a crowd of voyeurs, other puffins who gather around to watch what's going on and may even be motivated to start billing as well.

Storks also use billing to affirm their relationships. White Storks *Ciconia ciconia* breed in Europe and mostly overwinter in sub-Saharan Africa, forming monogamous pairs for the duration of the breeding season. It was traditionally believed that they paired for life, but it is now known that pairs usually migrate and overwinter separately and that any reforming of the bond between individual birds each spring is as much to do with their loyalty to a particular site as to each other. Although nests may be used for many years in succession, they may not necessarily involve the same pair or its generations of young. Male storks usually return to the nest site a few days ahead of the females. They may start gathering material to renew the nest, but the real action begins when their regular mate – or another female – arrives. A series of greeting ceremonies ensues, in which both birds throw their heads backwards onto their backs and clatter their mandibles together to create a machine gun-style rattling sound. Both birds may also undertake short display flights around the nesting area and develop their bond through billing.

The behaviour of birds in reaction to the demise of a mate is an area of increasing research. In seasonally monogamous species such an event is usually quickly overcome and the search for a replacement can begin almost immediately if eggs have not yet been laid. If, however, incubation has started or there are young in the nest, the remaining partner may try to raise the clutch on his or her own. Success rates vary, but brood failure is the usual outcome in cases where one bird is attempting to achieve what usually takes two. In species that pair for life the reaction to a dead or missing partner is much more profound. Many such birds are long-lived, so may remain with the same mate constantly for 20 years or more. The pair bond is thus very close and its rupture can be met with discernible distress and followed by subdued behaviour and what may even be regarded as a period of mourning. The reaction is especially intense if a dead partner is visible, but appears to subside if the body is removed. Albatrosses in particular appear to grieve for several months before showing interest in finding a replacement mate by resuming courtship display.

Whilst it is important to distinguish between genuine avian responses and anthropomorphic speculation, there is some evidence that certain birds have a sense of what we might term 'moral intelligence' and are indeed capable of expressing emotion. Observation of members of the crow family, such as magpies *Pica* spp., has revealed intriguing behaviour. This has included incidents of several magpies gathering around one of their dead peers, pecking gently at the corpse as if to revive it and even bringing plant material and laying it next to, or on, the body. Such reactions certainly suggest that bonds between birds may exist on a deeper level than has hitherto been understood.

Opposite above White Storks display to each other when first returning to their nesting site. Opposite below Vultures usually mate for life and are site-faithful. Even migrant species such as these Egyptian Vultures *Neophron percnopterus* usually return to the same breeding grounds.

COPING WITH POLYGAMY

Whilst the majority of birds are monogamous during the breeding season, 10 per cent or so are regularly polygamous. These include most of the species that lek, as well as a host of other birds from a wide range of families. In most cases the male bird mates with two or more females, an arrangement called polygyny, but in some species a female takes more than one mate in what is known as polyandry. The situation is far from straightforward, however, with some species also practising monogamy in certain situations and others – such as the Dunnock *Prunella modularis* (see page 94) – having highly complex mating systems that also embrace polygyny, polyandry and even polygynandry, in which two or more males 'pair' with two or more females.

Under the umbrella term of polygamy there are various such permutations. The males of some species engage in what can be termed harem polygyny, whereby they maintain several females, while others demonstrate what could be called serial polygyny or promiscuous behaviour, mating opportunistically with random females with which they then have no further contact. In the latter case the males play no role in egg incubation or the rearing of young, and the apparent advantages to the male of this system are clear: for minimal effort – amounting to little more than some courtship display and the act of mating – a male is able to maximize his genetic potential. However, the success of this strategy depends on various factors, not least the ability of the female to rear a family single-handedly. For this reason most polygamous birds have chicks that can either leave the nest shortly after hatching and fend for themselves with the minimum of parental supervision or care – termed nidifugous (see page 142) – or are fed by the female (or male, in the case of polyandrous species) alone on a plentiful and readily accessible food source.

Males that are serial polygynists do not bother defending any female with which they have mated, in the way that monogamous males might do. Free from the responsibilities of parenthood, such males are able to dedicate much of their time and energy to vigorous displays lasting several weeks that have a life and momentum of their own, irrespective of the responsibilities of the females with which they may have mated. Nowhere is this more keenly apparent than with lekking gamebirds (see page 52). Female birds visit the lek to find a suitable male with which to mate and, once the brief act of coitus is over, retire from the proceedings to set about finding a site for their nest. They have nothing more to do with the male birds, many of which do not always succeed in mating every year anyway. Young males in particular must 'wait their turn', as it is usually a handful of mature alphas who carry out the overwhelming majority of couplings with visiting females. Such promiscuity is unusual in the bird world, but not unique. Hummingbirds, for example, engage in a similar sexual free-for-all that sees male and female birds leading essentially solitary lives for much of the year and only coming together briefly to mate.

However, for polygynous males with harems the situation is rather different. Western Marsh Harriers, for example, are usually monogamous, but in areas where the population is particularly concentrated or there is an imbalance of sexes, a male may mate with two or even three females in one season and end up with multiple nests within his territory. He will usually endeavour to help build the nests, at least to the point of gathering potential

Greater Sage Grouse males busy at the lek. Dawn is the best time to observe peak levels of activity, although lekking continues on and off all day.

ROLE REVERSAL

Less than 2 per cent of bird species are polyandrous, with the female mating with two or more males in any one breeding season and usually playing little or no role in the rearing of young. In such cases she will often defend a territory, court a male or males (to the extent of actively competing with other females, then defending her 'prize'), and lay her eggs in a nest – built by the male – before leaving him to incubate the eggs while she turns her attention to courting more males. In this way polyandrous females of species such as the Spotted Sandpiper *Actitis macularius* may lay up to five clutches of four eggs every spring, each clutch fertilized by a different partner.

A similar arrangement applies in the case of phalaropes. In polyandrous species the female is often bigger and more brightly coloured than the male, a situation known as reverse sexual dimorphism. This is particularly obvious in phalaropes, both sexes of which undergo a pre-nuptial moult but where the female emerges with the brightest plumage. She then takes the lead in courtship and, after mating with as many as three or four males and laying clutches of her eggs in their nests, she heads off on her southwards migration. She may be thousands of kilometres away before any of her eggs even hatch.

The Greater Painted-snipe *Rostratula benghalensis* has a similar reproductive arrangement, but the difference between male and female plumages is much more marked than in phalaropes. The female is boldly marked and has a bright chestnut neck and chest, which on the male are a much more subdued grey-green hue – rendering him more inconspicuous when sitting on the nest.

One of the most notable polyandrous species is the Southern Cassowary *Casuarius casuarius*. Cassowaries are solitary for much of the year, but with the approach of the breeding season (usually June–October) the females become more tolerant of the males and eventually choose one with which to form a temporary partnership. They mate and remain together for a few weeks until the female is ready to lay her eggs. The male will have prepared a 'nest', little more than a scrape in the rainforest floor, lined with a few leaves. Once the eggs are laid, the female leaves the area and the male begins incubation. The female may then go on to mate with a second male and lay another clutch of eggs.

In only a handful of polyandrous species does the female maintain several male partners simultaneously. The Northern Jacana *Jacana spinosa* is characteristically known for this behaviour, with a single female maintaining a large territory within which she mates with several consorts and lays eggs in their various nests. Simultaneous polyandry is also recorded among Dunnocks, but the breeding habits of this species cover virtually all possible options and in the human context would be regarded as verging on the scandalous.

In the reverse dimorphic world of the **Grey Phalarope** *Phalaropus fulicarius* brightly coloured females vie for the attentions of the dowdier male.

93

A DARKER SIDE

Whilst some pairs of Dunnocks are monogamous, polygamy is generally the norm for this species, with 'pairs' comprising various arrangements of males and females, including trios where two males are paired with one female. In such cases each male may mate with the female as much as 100 times in a day. Only in the 1990s was it discovered that Dunnock males have a strategy for ensuring that, despite the obvious competition created by this situation, their genes may still triumph. When a male approaches a female and she exposes her cloaca, he will peck at it to force the expulsion of semen from any previous coupling with another male. Once he is satisfied that the semen is ejected, he mates with her, replacing the earlier male's genetic material with his own. Paradoxically, there is also a cooperative aspect to the Dunnock's sexual lifestyle, with two or more male birds sometimes helping with the care of the young of one female with which they have mated, even though the exact parentage of the chicks must be unclear.

Other birds may not be so forgiving. Even with essentially monogamous species, infanticide is not unknown, especially when one male is seeking to replace another. Ospreys *Pandion haliaetus* are notorious for such behaviour. At the famous breeding site of Loch Garten in Scotland, there have been incidents when an established male Osprey has returned in spring to find his regular female already paired with a new male and even sitting on eggs in the traditional nest. The late arriver has then been known to kick the usurping male's eggs out of the nest and reassert his position by mating with the female, then guarding her against the usurper. Birds such as Barn Swallows *Hirundo rustica* have even been recorded killing nestlings, with male birds – presumably unmated individuals – removing young from a nest, invariably in cases where the original male has disappeared, before mating with the hijacked female and building a new nest for her to lay a fresh clutch of eggs in.

Above However unassuming in appearance, **Dunnocks** have complex sex lives and a ruthless approach to genetic transmission. **Opposite** The urge to procreate is so powerful that **Osprey** males 'take over' females from weaker rivals and mate repeatedly with them so their sperm fertilizes the eggs.

PARASITISM

Brood parasitism, or the laying of eggs by one species in the nests of other species, is only recorded regularly in five bird families. The best known of these are the cuckoos (Cuculidae) of Eurasia and Africa, and the cowbirds *Molothrus* spp. of the Americas. In Africa, whydahs and honeyguides also parasitize other birds, as does one species of wildfowl, the Black-headed Duck *Heteronetta atricapilla* of South America. Other birds have occasionally been recorded as doing so, but more common is egg-laying in the nests of other birds of the same species (see page 130).

The popular perception of brood parasitism is that the birds concerned never rear their own young. This is certainly the case with the majority of such species, including the Common Cuckoo *Cuculus canorus*, a summer visitor to Eurasia from its winter quarters in Africa. More than 50 bird species have been recorded as parasitized by Common Cuckoos, with ten or so species providing the overwhelming majority of host nests. Female cuckoos specialize in favouring a particular host species and, with avoidance of detection so important, have evolved genetically to lay eggs that broadly match those of their host in both coloration and general pattern. They appear to inherit this specialism from their mothers.

When ready to lay, the female cuckoo locates a nest of the appropriate host species and while it is unattended, lands on it, removes one of the host's eggs and lays one of her own to replace it. The entire process takes only a few seconds, and the host birds return to their nest none the wiser. The system only works, however, if the host eggs are freshly laid – otherwise the cuckoo egg will hatch too late and be at a disadvantage, so in cases where the female cuckoo detects that the host eggs are too far advanced, she will remove them all, thereby forcing the hosts to lay a new clutch that will be better synchronized with her own egg (see page 146 for hatchling behaviour).

Female cuckoos may lay up to 40 eggs in any one breeding season, but perhaps this is not quite so remarkable given the energy savings the birds make in terms of neither building a nest nor incubating their eggs or feeding their young.

Egg mimicry is also found among other cuckoos, as well as in certain cowbird species, although not in the Brown-headed Cowbird *Molothrus ater*, which has been recorded parasitizing over 200 different species. Far less discriminatory and specialized than the Common Cuckoo, for example, it is not capable of egg mimicry and, as a result, its eggs and even its young are sometimes thrown out of the nest by the foster parents.

Meanwhile, not all cuckoos are serial brood parasites. The Yellow-billed and Black-billed species – *Coccyzus americanus* and *C. erythropthalmus* respectively – usually make their own nests and rear their own young, only occasionally resorting to a host. Even within parasitic families there are oddities. The Bay-winged Cowbird *Molothrus badius* is itself a victim of brood parasitism by other cowbird species, for example. It does, however, engage in a primitive form of parasitic behaviour in that it will often use the old nest of another species in which to lay its eggs, rather than building its own. Such appropriation of nests is also known among birds of prey, and there are plentiful examples of species that simultaneously share nests on a regular basis.

Female **Common Cuckoos** are dependent on a specific host species and must therefore seek out the relevant nests in which to lay their eggs.

Making nests

With pairs formed and mating taking place, for most birds constructing a nest is the next stage in the reproductive process. In its crudest sense, a nest is a structure in which eggs are laid and where chicks may develop once hatched. The provision of shelter, warmth and safety are thus prerequisite, with birds achieving these conditions via a bewildering array of different nest types and designs.

The range of nest-making material is equally vast, and the versatility of birds is nowhere more apparent than in their opportunistic choice of nest locations. Almost any conceivable situation may be used, from traditional sites in trees or rock crevices to contemporary ones such as the roofs of skyscrapers or even within the workings of operational machinery. Yet there are also many species that either do not build nests or prefer to use the disused structures of other species, as well as others that do not incubate their own young, relying instead on other birds or even on extraneous heat sources.

Yellow-rumped Caciques *Cacicus cela* weave hanging nests in trees, often near bee or wasp nests for protection from predators.

THE MINIMALIST APPROACH

Some birds do not bother to make any form of nest at all. They lay their eggs directly on the ground or another surface, such as a rocky ledge or even a tree branch. In such exposed environments the eggs are especially vulnerable to mishap and predation, so require careful attention from the parents at all times. Their vigilance is reinforced by evolutionary safeguards such as cryptic egg coloration (see page 128).

Open-country birds such as coursers (family Glariolidae) lay their eggs in a slight depression in the ground. They may enhance the hollow by scraping out soil with their feet, then rounding it into a saucer-like shape with their chest by crouching and rotating their body, but that is the limit of their nest-making effort. There is no nest material and, other than camouflage, no form of protection for the eggs or chicks from the elements or other potential threats bar the presence of the parents. Ground-nesters sometimes lay

their eggs in the lee of a rock, tussock of grass or fallen branch, or even in the hoofprint of a large mammal, all of which may afford some shelter, but it is equally common to find birds incubating their eggs right out in the open. Many species of tern are as economical in terms of nest creation, but they usually nest colonially and can therefore rely on group action to drive away would-be predators.

A similarly spare approach to nest-making is taken by cliff-breeding seabirds such as guillemots and Razorbills *Alca torda*. These birds nest colonially in concentrations that can involve hundreds of thousands of birds, all jostling for position cheek-by-jowl on precipitous terrain with little margin for error. They lay their eggs – one is the norm – on bare ledges that are often only a few centimetres wide and sometimes pitched at a perilous angle.

Yet as precarious as nesting on cliff faces might be, cliffs are perhaps not the riskiest place on which to lay eggs. White Terns often deposit their eggs on thin tree branches, making no attempt at a nest and with the female simply laying her single egg in an indentation in the branch or in its crook. This behaviour is highly unusual among terns and may have evolved in response to the presence of ground-based parasites or predators. Even so, it is a fraught strategy, with high casualty rates incurred during strong winds and heavy rain. The females lay again rapidly following the loss of an egg, so resuming the reproductive process with minimal delay – an example of an evolutionary response to specific recurring conditions.

Opposite This **Temminck's Courser** *Cursorius temminckii* is nesting in a classic open situation, its eggs laid in a shallow depression in the ground. **Above** The conical shape of **Razorbill** *Alca torda* eggs is an evolutionary adaptation that helps prevent them rolling off narrow cliff ledges.

Overleaf There can be fewer more precarious nest sites than the narrow branch often selected by **White Terns** on which to lay their single egg.

OPEN NESTS

Most species of tern, as well as many wading birds, gulls, gamebirds and nightjars, are ground-nesters, and whilst some lay their eggs directly on soil or shingle, others make a rudimentary nest. This is usually little more than a rough assemblage of pieces of grass, twigs or even small pebbles, used to create a basic platform. Nests at or on ground level are naturally susceptible to flooding, and in such cases birds are sometimes observed building up their nests with extra material in an attempt to keep eggs or chicks safe and dry.

With a few notable exceptions (see page 108) most duck and goose species make open nests on the ground, often hidden in low herbage or waterside vegetation. These usually comprise a shallow platform of grass, twigs or straw, lined with down plucked by the female bird from her own breast. The most iconic example is the Common Eider *Somateria mollissima*, but many waterfowl follow this pattern. They build their nest in a characteristic way, assembling a few rudimentary pieces of material and settling on them, before reaching out from this sitting position to gather other items, which they then tuck underneath and around their body to complete the arrangement.

In certain situations laying eggs directly on the ground – or even into a basic platform nest – is not a sensible option. This is clearly the case in the frozen wastes of Antarctica, where the Emperor Penguin *Aptenodytes forsteri* breeds in temperatures that can fall as low as −40°C (−40°F). The reproductive cycle of this iconic species is one of the most remarkable in the bird world and runs to a timescale that has evolved to ensure that chicks are fully fledged before the ferocious Antarctic winter begins each March. Because of the lengthy incubation and fledging period the year's breeding cycle actually begins the previous winter, in April, when male and female birds return to their traditional breeding sites some 100km

(60 miles) from the coast. They either relocate their partner from the previous year or find a new one, and following copulation the female lays a single egg, which she transfers immediately to the male, who is responsible for incubation. The transfer is made onto the top of the male's feet and thence into the so-called brood chamber, a flap of skin between his legs, with great care taken throughout to ensure that the egg does not come into contact with the frozen ground.

Some waterbird and tern species nest on aquatic vegetation on open water. Although this location has obvious advantages in terms of providing security from land-based predators, it requires constant vigilance. Water levels can rise suddenly after heavy rain and nests may need emergency work to ensure they do not become flooded. Whiskered Terns *Chlidonias hybridus* build shallow platforms on floating vegetation such as waterlilies, so are relatively immune to changes in water level. However, grebes often build their nesting platforms at the water's edge or on fixed structures such as underwater roots. When the water rises, they must raise the platform by adding material to it so that their eggs or chicks remain above the waterline. In heavy downpours this work may have to be carried out at considerable speed in order to avoid catastrophe.

Whiskered Terns build their nest platforms on floating vegetation on shallow lakes. Several thousand pairs may nest together in favoured locations.

TREE PLATFORMS

The rudimentary open platforms assembled by birds such as gulls are the basic template for the larger structures constructed by bigger ground-nesting birds such as cranes and swans. Similar nests are built by a variety of birds that erect their nests off the ground. These include most species of stork and heron, as well as birds of prey such as buzzards, eagles and vultures, which nest in large trees, on rocky ledges or even on electricity pylons. The principle behind the construction of an eagle nest in a tree is the same as that for, say, a gull on the ground, simply on a different scale: larger pieces of vegetation are arranged to provide stability at the base of the nest, with smaller elements used to build it up, then finer material employed to line the shallow depression on the surface in which the eggs and chicks sit.

Large birds such as storks and eagles assemble their nests mainly from twigs and branches, which help to anchor what often become very substantial structures that are used year after year and simply upgraded at the start of each breeding season. Such nests may be occupied every year for decades, even centuries – one case in Germany involved a White Stork nest that was first recorded in 1549 and was still in use in 1930. Raptor nests – often known as eyries, especially in the case of the larger birds of prey such as eagles – are also used regularly by successive generations of birds. With new material added each year, such nests can become huge, their weight even contributing to their demise in cases where the tree branches on which they are sitting can no longer withstand the weight and snap.

Not all platform nests are so robust. Most pigeons and doves build notoriously flimsy affairs, seemingly haphazard arrangements of twigs that look incapable of withstanding heavy rain or high winds and yet usually get the job done. At the other end of the spectrum is one of the most remarkable platform-based nests of all, that of the Hamerkop *Scopus umbretta*. This enigmatic bird takes several weeks to construct an enormous edifice, often comprising several thousand sticks and usually located in the fork of a mature tree (see page 125). The massive platform is not open, however, instead supporting walls and a dome-like roof, all held together by mud and with access provided via a tunnel leading to a breeding chamber within. The nest is usually occupied year after year by the adult pair, and is also often inhabited by various other types of wildlife, including bees, lizards and mongooses. Disused Hamerkop nests are frequently taken over by other birds such as owls and kestrels, and can even provide resting places for much larger beasts, including leopards.

Above Bald Eagle *Haliaeetus leurocephalus* nests reach huge sizes and are refreshed each breeding season. **Opposite** The **Jabiru** *Jabiru mycteria* builds massive platform nests that are used by the same pair year after year.

CAVITY NESTERS

About half the world's bird families contain species that nest in cavities. These include birds that only nest in such situations and actually create the holes themselves – woodpeckers and kingfishers, for example – but also many other species that will make opportunistic use of a natural cavity, or of one created by another bird species. From small passerines such as tits and flycatchers to birds as large as hornbills and owls, cavities provide a potentially secure and weatherproof breeding environment and are therefore highly prized. Competition for such spaces is often intense, with territorial squabbles and evictions commonplace.

Cavities come in all shapes, sizes and locations, but each species generally has a preference. Tree holes are among the most obvious opportunities, and these can vary from a narrow space behind flaking bark – used by small passerines and specialists such as treecreepers – to tunnels with a nesting chamber at the end, which are characteristic of woodpeckers. Different species have subtly different requirements, so it is usually possible to tell from the location and size of an entrance hole in a tree which species created it. Although woodpeckers are highly territorial, their varied needs are such in areas of prime habitat that it is often possible to find several species coexisting without any apparent competition.

Most woodpecker excavation work is carried out in early spring, although new holes may be created at any time of year and can serve as roosts during the non-breeding season. The majority of woodpeckers excavate in dead or dying trees, with only the larger species being powerful enough to chip out cavities in the dense wood of a living tree. Tree species appears to be less important to a woodpecker than the wood's condition, and often a bird makes several holes before deciding the one to use for nesting. This may be due to the need to identify a tree with the most easily excavated rotten heartwood, which can be located through 'sounding out' the trunk or branch by tapping it. It can take a woodpecker two to four weeks to excavate a nest hole, which is generally positioned as high as possible to help deter predators. The entrance is scarcely larger than the bird itself, and gives onto a short tunnel that drops into the nesting chamber. There is no attempt at building a nest, and the eggs are laid on a simple bed of wood chippings.

Their very specific requirements would seem to preclude woodpeckers living anywhere other than in wooded areas, but this is not the case. The deserts of the southwestern United States are home to two specialized species, the Gila Woodpecker *Melanerpes uropygialis* (see page 5) and Gilded Flicker *Colaptes chrysoides*, both of which excavate nest holes in the trunks and limbs of large Saguaro cacti. As in more conventional members of their clan, the nest holes of these two species are eagerly coveted by others. Woodpeckers can even find themselves chased out of their would-be residences by other species immediately after they have completed the excavation, and abandoned holes are soon appropriated by opportunists drawn from a wide range of species. These include owls, as well as more unlikely birds – tree-nesting ducks like the Common Goldeneye, Bufflehead *Bucephala albeola* and Smew *Mergellus albellus*, for example – all of which rely on tree cavities, either natural or made by larger woodpeckers such as the Black *Dryocopus martius* or Pileated *D. pileatus*.

In most cases the new tenants of an old woodpecker hole must make do with the space and configuration provided by its creators,

Opposite Black Woodpeckers are among the largest and noisiest members of their family, and are especially conspicuous when excavating their nest holes.

as they do not generally have the hardware required to enlarge or alter holes. However, adaptations are sometimes possible around the entrance. The European Nuthatch *Sitta europaea* reduces the size of an entrance hole by plastering mud around it, thereby preventing Common Starlings *Sturnus vulgaris* – also frequent occupants of old woodpecker holes – from gaining access and assuming ownership.

Large natural holes in trees are often in short supply, and for sizeable birds like hornbills (family Bucerotidae) it can be especially difficult to find suitable nesting sites. Rotting cavities or the hollowed out stumps of large branches broken off in storms offer the best opportunities for what is one of the most singular nesting strategies in the whole bird world. When the female hornbill is preparing to lay her eggs, she enters the nest hole; both adult birds then plaster up the entrance, using the female's droppings, until only a narrow aperture is left. The female remains sealed inside, with the male feeding her via the small opening, all through incubation, only breaking open the seal when the young are ready to leave the nest. This form of temporary confinement is unique in bird terms and is probably an extreme adaptation of a protective strategy for the eggs and chicks.

Rock cavities, whether natural or man-made – such as stone walls or crevices in buildings – provide relatively secure breeding environments for birds such as sparrows, wrens, swifts, starlings, wheatears, jackdaws and certain types of owl and dove, as well as more unusual species. Black Guillemots *Cephus grylle* and storm-petrels (families Oceanitinae and Hydrobatinae), for example, often utilize crevices in rocks, scree and even tumbledown buildings. Natural cavities in or on the ground, for example among or under tree roots or in old termite mounds, are also popular with certain species as breeding locations. Iconic New Zealand birds such as kiwis *Apteryx* spp. and the critically endangered Kakapo *Strigops habroptila* nest on the ground, often in hollow tree trunks. With no indigenous land-based predators in New Zealand, such sites would have been relatively safe historically. However, the 19th-century introduction of the non-native Stoat and domestic cat have had a disastrous impact.

Few birds actually dig their own burrows in the ground, preferring instead to occupy those originally excavated by other animals. Common Shelducks, for example, often use abandoned rabbit holes, as do puffins, which will even attempt to take control of occupied spaces by evicting the rabbits. Burrowing Owls *Athene cunicularia* meanwhile favour burrows dug by rodents such as Ground Squirrels and Prairie Dogs (and are often in competition with those species for accommodation), but – like puffins – they excavate their own nesting holes if necessary and the local ground conditions allow, using their feet to scrape away the soil. One of the most remarkable burrows excavated by a bird is that of the diminutive Rhinoceros Auklet *Cerorhinca monocerata*, which breeds colonially on the islands and coastline around the North Pacific. At sites where there is a paucity of suitable natural cavities, both parents use their bills and long claws to excavate a nesting burrow up to 6m (20ft) in length. By digging to such a depth, and by only visiting their burrows during the night, the auklets minimize the risk of predation by gulls.

Birds such as kingfishers and bee-eaters excavate their own nesting sites within soft substrates on vertical and exposed banks. The birds start excavating by flying repeatedly up to a bank and digging at the soil with their bills, usually whilst hovering briefly. Once they have made an initial indentation and obtained a foothold, they use their bills and feet to dig further into the bank, creating first a tunnel then a nesting chamber at the end. In some cases the tunnel can reach up to 6m (20ft) in length and in colonial situations, where many pairs may have repeatedly excavated in close proximity to one another, this activity can destabilize a bank and ultimately contribute to its collapse.

Above **Burrowing Owls** at the entrance to their nest hole, which is often a disused rodent burrow.

Overleaf A male **Great Hornbill** *Buceros bicornis* at a nest hole. His mate and young are incarcerated inside and will not emerge until the chicks are ready to fledge.

As in the case of birds that use tree cavities, the precise arrangement within a ground or bank nesting hole or chamber varies according to the species. Most passerines make an actual nest within the cavity, usually of twigs and lined with softer matter such as moss or feathers, whilst other birds may assemble a basic layer of material or simply lay their eggs on the bare surface within the hole.

NESTBOXES

The provision by humans of artificial nesting sites for wild birds goes back thousands of years, to when certain species were 'managed' commercially for food and/or profit. Only in the 19th century did the idea take hold that watching birds use nextboxes could be a source of enjoyment and interest for its own sake. Since then, enthusiasm for providing potential breeding sites for birds has grown enormously, and today nestboxes are valued not only by the general public but also by scientists compiling data on bird populations. There are now almost five million nestboxes provided in British gardens alone, a development that has certainly helped sustain and even boost populations of birds such as tits. Worldwide, at least 150 bird species have been recorded breeding in nestboxes.

Tailor-designed boxes now exist for a wide variety of cavity-nesting birds, with varying degrees of success. Some species take readily to artificial structures, whilst others may only use them after a lengthy period of habituation or when no natural site is available. Nestboxes can have a particularly positive impact in habitats where the number of available natural cavities is declining or restricted, thereby reducing bird numbers or inhibiting population expansion. In eastern North America, boxes have been instrumental in helping change the fortunes of the Eastern Bluebird *Sialia sialis*, a species that during the 20th century was in

decline across much of its range due to the loss of suitable nesting habitat. It has been helped on a local level at least by the increasing provision of backyard boxes.

Artificial nesting platforms erected on the tops of poles or buildings have proved attractive to nesting White Storks, Ospreys and Peregrine Falcons *Falco peregrinus*, and when moored out in the middle of a lake or lagoon are regularly used by terns. In parts of central Europe Long-eared Owls *Asio otus* have a penchant for nesting in wicker baskets placed in trees, whilst in Scotland the population of Black-throated Divers *Gavia arctica* has benefited from the provision of floating breeding platforms, which rise and fall with changing water levels and are thereby more secure than natural sites, which are prone to flooding.

Above Like most members of their family, **Southern Carmine Bee-eaters** *Merops nubicus* nest colonially and usually breed at the same site each year.

Above **Eastern Bluebirds** take readily to nestboxes, a factor that has helped to arrest their decline in some parts of their range.

CUP NESTS

The vast majority of birds hide their nests in vegetation, with many choosing to do so in shrubs or trees, where the combination of height and foliage provides a degree of safety from predators. There are still risks of course, with plenty of predatory birds, reptiles and mammals able to reach nests in trees, and the constant threat of eggs and young toppling out, but generally speaking success rates for birds that nest arboreally are up to 30 per cent greater than for those nesting on the ground.

The most secure breeding conditions in trees are obtained through building a cup-shaped nest. This is the basic nest template followed by most passerines, including those that breed in low herbage on or near the ground, and the method of construction is broadly the same, even if the end results can look dramatically different. The bird begins by identifying a suitable location for the nest, usually a fork or a crotch in a branch. He or, more likely, she then makes some anchor points, often by looping strands of silk – taken from insect cocoons or spiders' webs – around the supporting branches. An outer shell of coarse material such as twigs is then assembled between the anchors, with the bird pushing many of the twigs towards a core central point so that they radiate out and touch adjoining branches and stems. This helps create structural pressure to hold the shell in place. A second layer of slightly finer padding material is laid down within the shell, and on top of this layer comes the final cup. This is formed by the bird rounding out the shape with its chest and lining it with finer pieces of vegetation such as moss and pieces of leaf, as well as feathers and/or animal hair.

The bird uses its bill to integrate and weave together the various nest elements, packing softer material between the framework to make the nest weatherproof and solid. Mud may be applied to give greater stability. Precisely what materials are used in nest-making depends largely on what is readily to hand, although some birds specialize in using particular media: many hummingbirds, for example, build their nests almost exclusively from moss, lichen and spiders' webs, whilst the Eurasian Penduline Tit makes extensive use of plant down. Birds compete with each other for nest material, often stealing choice pieces from other nests or even out of the bills of neighbouring pairs.

Nest building is arduous work, and certain species of finch have been recorded making up to 1,000 individual trips to gather nest material for their nests during the construction period of a week or so. The exact level of task-sharing undertaken by male and female varies according to species, but in the case of passerines building cup nests it is usually the female that carries out the majority of the construction. The male may often start the nest, or even indicate the proposed site, with the female building it and usually carrying responsibility for the finishing touches, such as the cup lining. In cases where the female does the majority of the nest building she will usually also carry out most or all of the incubation. With non-passerines the nest-building and incubation tasks are often more equally shared between the sexes.

BEYOND THE CUP NEST

The cup-nest format has the advantage of being highly adaptable, with different species adding certain features to suit their particular requirements. Birds that nest high up in the tree canopy need to take precautions to secure their nest against high winds.

Opposite Thrushes such as this **Fieldfare** *Turdus pilaris* make typical cup nests. The inside will be lined with mud.

Certain species of oriole, for example, make a cup nest that is slung between two branches, rather like a hammock, and significantly deeper than that made by many other passerines. Tailorbirds *Orthotomus sutorius* stitch leaves together to make a pocket within which they construct their pouch-like nest, whilst species such as the Long-tailed Tit *Aegithalos caudatus* have developed the cup format into a ball-shaped dome with a roof and a side entrance, providing greater shelter and protection.

The dome format has also been adopted by the South American ovenbirds (members of the large family Furnariidae), which construct some of the most impressive examples of all avian architecture. Using mud strengthened with plant fibres and animal hair, they build dome-like edifices weighing up to 4kg (9lb), usually located in prominent positions such as on fence-posts or horizontal branches. Access is via a side entrance and, once baked by the sun, the result resembles a traditional clay oven, fairly secure from predators and with its almost impermeable walls providing insulation that helps keep temperatures inside relatively equable.

Mud is also a key component in the nests of birds like swallows, swifts and martins. In prehistoric times they would have been wholly cliff- and tree-nesting, but nowadays many species take advantage of man-made structures. Barn Swallows famously build their open-cup nests of mud, grass and other material, such as animal hair, on beams or ledges inside buildings, often wedged into the right angle against a vertical projection, while Common House Martins *Delichon urbicum* only nest on external walls (or cliff faces) under a sheltering overhang – most typically a roof or gable. They collect pellets of wet mud in their bills from the sides of rivers or lakes, or often from rain puddles, then return to the nest site, sticking the pellets onto the vertical wall and building up an initial crescent-shaped line that forms the outer profile of the nest. This process is continued until a cup-shaped structure has been created, with its 'ceiling' formed by the overhang and

with a single entrance at the upper edge. Swallow species such as the Red-rumped Swallow *Cecropis daurica* and American Cliff Swallow *Petrochelidon pyrrhonota* use similar techniques to construct flask-shaped nests which, in the case of the latter species, a colonial nester, frequently create a dramatic honeycomb effect as one nest is fixed ingeniously against another.

Although certain species of swift build little or nothing by way of a nest, instead laying their eggs in crevices in cliffs or buildings, others build more elaborate affairs and are notable for their use of saliva as a binding agent. Material for the nest, such as spider silk, feathers and pieces of grass and other plant material, is gathered in mid-air or plucked from trees or bushes. It is then worked into a cup, using the viscous saliva to glue everything together and to fix the nest to vertical surfaces such as walls and the interiors of chimneys or, in the case of the African Palm Swift *Cypsiurus parvus*, the undersides of palm leaves. The so-called edible-nest swiftlets *Aerodramus* spp. of south-east Asia, compose their nests almost entirely of dried saliva, with little by way of foreign matter. These are highly prized for 'bird's nest soup', with many millions of nests still harvested annually to meet the regional demand for this supposed delicacy.

Above **Common House Martins** gathering mud with which to construct their nests. In dry weather a lack of suitable material can inhibit their breeding success.

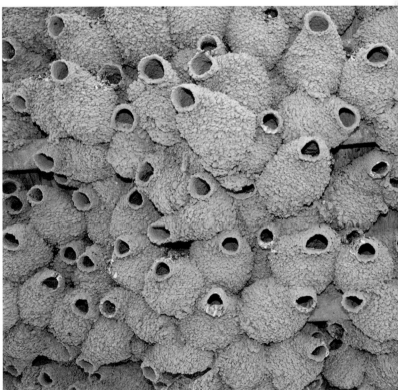

Above left A **Rufous Hornero** *Furnarius rufus*, a type of ovenbird, singing from atop its 'nest'. **Top right** **Eurasian Golden Orioles** *Oriolus oriolus* make a sling-type nest, usually strung between a fork in a branch. **Below right** Not all swallow species breed colonially, but over 3,000 **American Cliff Swallow** nests have been recorded in colonies.

ULTIMATE REFINEMENTS

Some birds have developed the core nest concept into something rather unusual and sophisticated in terms of design and construction. These include the New World orioles known as oropendolas *Psarocolius* spp., colonial nesters that build skifully woven pear-shaped nests. They are attached to tree branches and dangle down up to 2m (6.5ft) into mid-air, making them virtually predator-proof, if vulnerable to high winds.

It is, however, the weavers (Ploceidae) that are arguably the greatest masters of the art of nest-building. There are more than 50 species of true weaver and a tremendous range of nest types and construction techniques within the family. Most species are colonial, with weaver communities ranging from a few pairs to concentrations of 500 birds or more. The Sociable Weaver *Philetairus socius* in particular builds enormous structures of twigs, grass and plant fibre, which can end up swamping the trees or telegraph poles on which they are assembled.

These avian condominiums contain scores of self-contained nest chambers within the greater 'umbrella', which is started by birds – working together – stuffing pieces of grass into nooks and crannies in the tree bark or wood. A viable mass of grass and straw is assembled, then work starts on the roof, which is composed largely of twigs and is built up into a dome-like shape. Once it is complete and the substructure below is weatherproof, the birds begin to build the nest chambers within the mass of grass below, at first excavating a suitable space, then adding new material and weaving the chamber walls into shape. They then line these with finer material and construct a tunnel entrance that opens downwards. New chambers are added to the mass as and when required, which explains the amorphous and organic character of many established Sociable Weaver nests.

A typical form of weaver nest is a ball-like structure woven from plant fibre and built either between upright supports such as plant stems or suspended from a branch or other piece of vegetation.

The nest is started by making a bridge or necklace of fibre between two or more vertical anchor points or attached to a branch. The bird then sits on the bridge or necklace and adds more material until a pad and then a cup is formed. It constructs walls around the cup, and extends the rear wall upwards and over to form a roof, thereby completing the spherical structure. Access is via a narrow entrance, usually at the side but sometimes underneath via an extended tube.

Birds do not restrict themselves to natural materials when nest-building. Many species – especially those living in urban areas – incorporate into their nests an unlikely array of man-made items, from string, wire and ring-pulls from cans to pieces of newspaper, polythene and discarded human clothing. Crows are particularly keen on such additions, and have even been recorded building entire nests almost exclusively from lengths of plastic or metal cabling, or from wire coat-hangers.

Black Kites *Milvus migrans* in the Coto Doñana National Park in Spain have been recorded lining their nests with strips of white plastic. A survey revealed that 77 per cent of these nests contained additional material of this type, with birds in peak breeding condition and occupying the best nesting sites most likely to include it. Pairs with nests containing white plastic tended to raise more chicks successfully, although exactly why remains unclear. The plastic may act as a territorial signal and a sign of genetic strength, or it may serve to camouflage the kites' white eggs and make it harder for predators such as crows to see them.

Opposite A male **Red-headed Weaver** *Anaplectes rubriceps* constructing its nest. Learning all the refinements comes with age and experience.

For some species, simply building a nest is not enough. Once it is complete, they attempt to decorate it, adorning it with objects both natural and man-made, other than those used in its construction. It has been known for centuries that members of the crow family, particularly magpies, have a penchant for taking unusual objects to their nests. Some birds of prey do this also, with Red Kites *Milvus milvus* notorious for 'decorating' their nests with foreign objects that play no practical role in the nest's structure. Such accoutrements range from old Christmas decorations and pieces of coloured glass, to children's toys and even flowers stolen from graves in cemeteries.

Increasing numbers of bird species are now understood to incorporate aromatic plants into their nests. Herbs such as mint, yarrow and lavender are thought to be used by birds in this way as a means of maintaining or improving nest hygiene, because they contain chemicals and bacterial properties that help keep pest numbers down. Passerines were traditionally not thought to have a defined sense of smell, but this opinion is now under revision. Male Common Starlings have been recorded selecting aromatic vegetation to take back to their mate on the nest, possibly as a stimulus to her to lay more eggs.

Above Icons of the African landscape, **Sociable Weaver** nests are among the most remarkable constructions made by any creature. **Opposite** **Black Kite** chicks hatching out in a nest containing old rags and other extraneous matter brought to the nest by the parents.

TAKING ADVANTAGE

Brood parasites such as cuckoos and cowbirds neither make a nest nor rear their own young (see page 96). Some other birds manage the latter but forego the former, preferring to occupy a disused nest originally built by an entirely different species. Raptors and owls do this as a matter of course, with Hobbies *Falco subbuteo* and Lanner Falcons *F. biarmicus* regularly using the old nests of corvids, and Great Horned Owls *Bubo virginianus* taking over that of a hawk or even a heron.

The benefits of association lead certain birds to deliberately locate their nests close to those of other species. This behaviour is usually motivated by the advantages that such arrangements may bring in terms of protection, with small birds often consorting with larger passerines that may challenge potential predators more robustly than would be possible for more diminutive species. Equally, ducks nest regularly in colonies of terns, relying on the latter's collective mobbing behaviour to keep them safe from gulls looking for eggs and chicks. Weavers seek out the vicarious protection of buzzards and eagles by nesting near their eyries, thereby benefiting from the deterrent effect of the raptors' presence on potential weaver predators such as snakes and monkeys. Other aggressive creatures, such as bees and wasps, may also find themselves used in this way, as do humans – birds nesting close to human habitation often show improved breeding success over those living in more traditional environments.

Meanwhile, very large nests – such as those constructed by storks – regularly support breeding colonies of smaller birds like sparrows, which fashion their nest chambers in the understorey of the larger structure. Even jackdaws and doves choose to nest in such situations. This behaviour may simply be opportunistic, but hanging out with the big birds clearly has its benefits in the battle to rear young successfully.

Above A female **Common Blackbird** on her nest in a functioning traffic light. The unusual location offers warmth and protection from the elements.
Opposite **Hamerkop** nests can be enormous and provide accommodation to a range of wildlife in addition to their original makers.

Eggs and young

An appropriate nesting site, whether it be a simple depression in the ground or a complex woven pouch high up in a tree, is the immediate starting point for the process of egg-laying and incubation. Natural selection has helped determine how many eggs each species lays in a clutch, and how many clutches are attempted within each breeding season. These numbers are governed by factors such as where particular birds live, the nature and abundance of their food supply and the longevity of the species concerned, and are geared around the production of the maximum number of young that the parents are able to rear to independence.

The extent of parental duties is equally varied and in terms of duration can last from barely a month (many passerines) to more than a year (the larger albatrosses) – or, in the case of brood parasites, it takes up no time at all, as responsibility is totally forsaken in favour of imposing upon a host species.

These **Little Ringed Plover** *Charadrius dubius* eggs are impeccably camouflaged against their shingle backdrop.

TIME TO LAY

Exactly when to lay, and how many eggs, is determined by various considerations, not least natural selection. Birds in tropical parts of the world tend to lay fewer eggs than those in more temperate zones, where due to the longer day length during spring and summer there is more time for the parent birds to gather food for their chicks, and therefore rear more. The number of clutches per season, and the total of eggs in each, therefore tends to increase with latitude, with birds such as warblers and thrushes laying up to 75 per cent more eggs per clutch in Scandinavia, for example, than they do in Spain.

Egg-laying is timed to coincide with a dependable and adequate food supply being available for the hatchlings, a factor that also helps determine clutch size. Eleonora's Falcon *Falco eleonorae*, a summer migrant to the Mediterranean, breeds unusually late in the year, timing the hatching of its eggs to coincide with the peak season for passerine migration southwards through the region, thereby helping to ensure an abundant supply of the small birds on which the falcons feed their chicks. Many species are able to regulate their breeding on a year-by-year basis, with birds of prey in particular laying larger clutches when food is plentiful. Short-eared Owls, for example, are heavily dependent on voles as prey, and may not breed at all in years when vole numbers are very low.

Small birds usually lay one egg a day until their clutch is complete, but in larger species the process is more prolonged, with one egg laid every few days. This naturally results in youngsters of different ages and sizes, a factor that leads to sibling competition (see page 146). Generally speaking, larger birds lay fewer eggs than smaller birds – for example, an albatross usually only lays one egg every other season, whilst a European Blue Tit may produce two clutches annually of up to 15 eggs each. Both strategies are designed to produce the end result of a sustainable population: albatrosses are long-lived birds with few natural predators, so they do not rely on rearing a young bird every year; tits, on the other hand, have an average life expectancy of less than two years and are predated by a wide variety of species, losing the majority of each year's new generations within a few weeks or months.

The precise number of eggs laid by a bird in any one season also depends on immediate success rates. Most species are able to lay another clutch if the first falls to predators or the weather, whereas some produce a fixed number of eggs per season, so cannot attempt to compensate for any losses and must effectively abandon their breeding attempt for that year. Brood parasites such as cuckoos and cowbirds may lay many eggs in one season in nests of their various hosts (see page 96), while others are part-time parasites of their own species. A female Common Moorhen *Gallinula chloropus* may lay a few eggs, seemingly at random, in nests belonging to other moorhens before settling down to incubate a full clutch in her own nest. In evolutionary terms this may represent a stage in the process towards brood parasitism.

Above The breeding success of **Short-eared Owls** *Asio flammeus* is closely tied to prey availability. In good years a pair may rear as many as eight or nine young.

Overleaf **Black-browed Albatrosses** *Thalassarche melanophrys* have one of the longest breeding cycles of any bird and only ever raise one chick at a time.

131

INCUBATION

Eggs must be kept at a particular temperature so that embryos develop at the appropriate rate. For the overwhelming majority of birds this is achieved via incubation, the use of body heat to regulate conditions in the nest and stimulate embryo growth to the point of hatching. Hormonal changes in the adult bird prompt the temporary development of an abdominal brood patch, an area of bare skin partly exposed by the shedding of a small area of feathers; this facilitates the transfer of heat from body to eggs via a mesh of blood vessels. The size and shape of the brood patch varies between species, with some birds having a centrally positioned single patch, but others — notably gulls — with a more extensive arrangement of three patches: one for each egg (three eggs being a typical clutch for gulls).

When settling down to incubate, a bird raises its breast and flank feathers to reveal the brood patch and shuffles into position over the eggs, ensuring they are neatly in position and nestling against the exposed skin. Brood patches appear in both male and female in species where both parents incubate; otherwise, only in the parent responsible for incubation. They disappear soon after incubation and a few bird species do not develop them at all. Gannets and pelicans, for example, transfer body heat to their eggs by cradling them in their feet.

The precise intensity and duration of incubation — which can last from ten days to as much as ten weeks — is dependent on individual species and local conditions. Whether it is the female, male or both parents that incubate the eggs is determined by species, with single-parent incubation usually the domain of polygamous birds. In many cases the duties are shared, and in some instances there is a distinct shift system in place; male birds may incubate the eggs at night, for example, whilst the female takes over during the day. In species in which the male is more brightly coloured, and therefore more obvious to predators when on the nest, such an arrangement has obvious advantages. Natural selection has helped ensure that in most cases the female's plumage is more effective as a camouflage. Incubation does not usually start until the last egg in a clutch is laid, thereby ensuring that all the embryos develop at the same rate and hatch more or less simultaneously. However, not all birds follow this pattern — raptors are a notable exception — and in cold climates, where leaving the first eggs unattended may expose them to chilling, incubation may also start before the clutch is complete, resulting in different hatching dates and varying sized youngsters.

Throughout incubation it is important that sudden temperature changes are avoided and that a steady level of warmth is maintained. Most eggs are incubated at between 34°C (93°F) and 37°C (98°F), somewhat less than the normal bird body temperature of 40°C (104°F). Birds therefore have to monitor conditions in the nest to ensure that their eggs do not overheat or become chilled. They do this via receptors on their brood patch, then adjust their brooding regime accordingly. Embryos are less sensitive to cold than to heat, so in very hot conditions parent birds often need to stand over the eggs to shade them from the sun and prevent them from cooking. Birds sit very still while incubating. This immobility both helps avoid detection and ensures that a steady temperature is maintained among the eggs below. However, every so often a

Opposite top This **Sooty Gull** *Larus hemprichii* is shading its eggs to prevent them overheating in the midday sun. **Opposite below** The obvious brood patch on this female **Snowy Owl** *Bubo scandiacus* is a clear sign that the bird is breeding.

sitting bird may turn its eggs, behaviour that is most common among species with large clutches. The brooding bird arches its neck downwards and drags its bill backwards through the eggs so that they roll over. Where the nest contains several eggs this action also serves to bring those on the periphery into the centre, so that all are evenly incubated.

Where both sexes play a role in incubation – and it is more usual for the female to carry the lion's share of the role – it is common for the non-incubating parent to bring food to the sitting bird. But when incubation is the responsibility of one parent alone, the eggs are necessarily left unattended while the bird goes off to feed. Its absence brings threats in terms of exposure to predators or unfavourable weather conditions. Risk levels are reduced by covering the eggs with nest material, which serves to hide them as well as helping to regulate excessive temperature change.

The Egyptian Plover *Pluvianus aegyptius* employs various strategies for keeping its eggs at the requisite temperature. This species breeds on exposed stretches of river or lake shorelines in tropical and sub-tropical Africa, where midday temperatures at surface ground level in the sun can reach 80°C (176°F). As the eggs are laid directly onto the sand, the plovers take care to prevent them from overheating by covering them with loose substrate to a depth of 10mm (0.4in). During the hottest part of the day they also continuously wet the eggs and/or the surrounding sand by soaking their ventral feathers in water, then settling on the 'nest'. This helps to keep the immediate temperature from reaching dangerously high levels. At cooler times, for example during the night, the parent birds may brood their eggs in the usual way, but they appear to be able to monitor changes in the nest temperature and take appropriate action when required.

One of the most remarkable incubations is that followed by the male Emperor Penguin, which incubates a single egg in a special

This **Brown Pelican** *Pelicanus occidentalis* is turning its eggs to ensure that each is maintained at an even temperature.

brood pouch (see page 104). Incubation takes 65 days or so, during which time the bird is unable to move to feed, relying instead on body fat laid down during the summer months. He loses as much as half his body weight during incubation, which coincides with the bitterest winter conditions, with the egg-brooding males huddling together to keep warm and protect their eggs. Meanwhile, the female birds have walked to the ocean to feed up, remaining there for several weeks before coming back to start feeding the newly hatched chicks with regurgitated fish. The males then leave for the coast, returning later to help with the feeding.

ALTERNATIVE INCUBATION

There are some birds – notably the megapodes (family Megapodiidae) of Australasia – that do not incubate their eggs with their own body heat at all, leaving that process to external elements such as the sun or the heat given off by decaying vegetation. This is, of course, the strategy employed by most reptiles, leading some biologists to concur with a popular notion that this group of birds represents a 'living link' with reptilians and even potentially with dinosaurs.

Megapodes are sometimes known as mound builders, due to the habit of some species of burying their eggs in piles of natural material. Overall, the family uses a variety of different means to harness natural heat as an incubator. The 'nests' can be as simple as a small pit dug in a patch of sunbaked sand and as monumental as a vast mound of soil and rotting plant matter reaching heights of 5m (16ft) and a diameter of 10m (33ft): the heaviest bird-made structures in the world.

Some species are highly specific in their needs. For example, the Micronesian Scrubfowl *Megapodius laperouse* relies on volcanic action to provide the heat required to incubate its eggs, by burying them in loose substrate near a source of geothermal steam.

Due to the requirement for such defined conditions, numbers of breeding megapodes are sometimes highly concentrated at especially favoured sites.

Forest-dwelling megapodes live largely in the shadows of the understorey, so cannot rely on the heat of the sun to incubate their eggs. It is largely these birds, therefore, that have evolved to construct the vast mounds of heat-generating decaying matter for which the family is renowned. These edifices are constructed by the male bird, which assembles a great heap of dead vegetation and soil – a compost heap, effectively – and encourages a female to lay her eggs on it. He then covers the clutch with more vegetation and soil.

Once buried in the fermenting pile, the eggs require a constant temperature of 33°C (91°F) to develop satisfactorily. The male is obliged to work constantly during daylight hours to ensure that the correct conditions are maintained. He does this by regularly testing the temperature – he inserts his bill into the mound and takes some of the soil or material, which he presses against his heat-sensitive palate. If the temperature within the mound is too high,

Above This **Maleo** *Macrocephalon maleo* pair is digging a pit in which the female will lay her eggs.

he removes some of the upper layers of vegetation so that the egg chamber beneath is cooled slightly. Conversely, he adds extra material if he detects that more heat is required.

This ability to adjust the temperature of the nest mound is vital for reproductive success, but megapodes have evolved other refinements that underpin their unique mode of breeding. Their eggs are among the largest of all birds' eggs in terms of their relative size to that of the bird that lays them. A single egg may amount to 16 per cent of the female's body weight and has a high yolk content, an essential element in the production of the highly advanced young megapode that will hatch out – in a virtually adult state – after two and a half months or so buried in the nest mound (see page 142).

Above The nest mound constructed by the male **Malleefowl** *Leipoa ocellata* is among the largest structures made by any bird.

HATCHING

The emergence of a chick from an egg is one of nature's great marvels and the culmination of a reproductive process than for most species has been going on for months – even years, in some cases. As the hatching date approaches, certain changes take place within the egg. The chick moves round inside so that its bill is facing the blunt end, where the air cell is located. Then, with a rapid movement of its head, it breaks the air cell and starts breathing with its lungs for the first time. At this point the chick may start calling, or 'cheeping', from within the egg, which alerts the parent to the imminence of hatching.

By this stage most chicks have also developed an egg tooth, a small bony protruberance on the tip of their bill. They use this to break a small hole in the shell to allow more air in – a process known as pipping – but over 24 hours may elapse before the next stage in the hatching process. The chick then works its way around inside the egg, usually anti-clockwise, chipping with its egg tooth to create a necklace of small cracks in the shell, usually towards the blunt end of the egg. A powerful muscle behind its head, the so-called hatching muscle, powers the blows it must make to break the shell, and the chick will pause to rest between its exertions. When the necklace is complete, it uses its head, body and limbs to push against the weakened sections, forcing the 'top' of the egg to detach and allowing it to emerge. The whole process can take from a few minutes to an hour or more.

Exhausted and panting from its efforts, the chick is now at its most vulnerable. The parents quickly remove the discarded pieces of eggshell – the bright white interior may catch a predator's eye – and start brooding the chick immediately so it dries off and does not chill. The egg tooth disappears within a few hours.

Above By using its head to force apart the weakened shell of an egg, a chick is able to emerge. **Opposite** A chick uses its egg tooth to break a small hole in an eggshell.

CHICKS

Exactly how developed or not a chick is upon hatching depends on the species. Most ground-nesting birds are well advanced when they hatch, covered in down, with their eyes open and with strong legs, enabling them to walk, run — and in some cases, swim — within minutes. They may leave the nest almost immediately and move around with their parents in search of food. Such hatchlings are known as precocial or nidifugous, the latter term referring to their early ability to leave the nest. Most waders, ducks, geese, swans and gamebirds come into this category. Nidifugous young rely on their parents heavily in the first few days, but such is their level of early independence that they are soon able to start looking for food themselves.

Some precocial young have to face surprising ordeals within hours of hatching. Tree-nesting ducks such as the Common Goldeneye, Wood Duck *Aix sponsa* and Goosander *Mergus merganser* must lead their young out of the nest hole and down to ground and thence to safety on the water. This process is fraught with danger, commencing with a seemingly death-defying leap of what can be 15m (50ft) or more from the nest down to the ground. One or both parents, usually the female, attempts to lead the way, leaving the nest hole and flapping about nearby in mid-air, noisily encouraging the ducklings to take the plunge. Understandably reluctant to do so, eventually one brave individual jumps out into the unknown, with the rest usually following suit.

So light are ducklings at that age that they simply float down unharmed to the ground, where the parents quickly round up their brood and lead them towards water. This stage of the journey is the most perilous, as there is little the adult birds can do if predators such as corvids, mustelids and foxes are about. If all goes well, the ducklings soon reach the relative security of the water, where they immediately take to swimming and can start to feed. A similarly arduous journey is made by young auks such as guillemots, which are encouraged by their parents to leap off their nesting ledges on precipitous cliff faces and flutter down — they are barely able to fly at this stage — to the sea below, where their parents await them. If they mistime their descent and land instead on a beach or among rocks, they are vulnerable to predators and must scramble quickly towards the water.

Young megapodes are the most precocious hatchlings of all and are equipped for a completely independent existence from the point of hatching. They emerge in a very advanced state of development, literally kicking their way out of their eggs with their feet, rather than using an egg tooth. They then tunnel their way out of their nest mound and are able to fly almost immediately.

At the other end of the spectrum are birds that hatch out blind and featherless, so are completely dependent on their parents. Such young are known as altricial or nidicolous, the latter term meaning 'in the nest', with most passerines coming under this heading. Their newly hatched chicks are completely helpless, cannot be left on their own for long and must be brooded as assiduously as eggs for the first few days. The parents' immediate concern is to bring food to the nest, a labour-intensive process that can involve them in several hundred separate trips per day. The instinctive reaction of the chicks to beg for food is triggered by the arrival of the parents at the nest, or by their calls. The chicks

Top Young wading birds such as these **American Oystercatchers**
Haematopus palliatus are able to walk and start searching for food soon
after hatching. Above **Goosanders** are tree-nesting birds and their young
often face a dangerous journey across open land before reaching water.

immediately respond by rearing up and opening their mouths, their brightly coloured gapes serving as both a signal to their parents to feed them and as a guide to where the food should go.

Fuelled by a high protein diet, the rate of growth is extremely fast. In small passerines such as tits and warblers, after two or three days tufts of downy feathers begin to appear on the chicks' heads in particular. Within a week feather tracts are forming on their backs and wings, where the feather sheaths that will eventually house their flight feathers are also starting to show. At nine or ten days the chicks are almost completely covered in young feathers, with few areas of naked skin remaining and feather tips emerging from the sheaths on the wings. At this age their eyes are also opening. By two weeks they are fully feathered and ready to leave the nest within a further week or so.

Nest sanitation is important for nidiculous birds, where a build-up of faeces – which are often white – and other debris may attract the attention of predators. Most such nestlings produce their faeces in a gelatinous sac, which the parents seize in their bills as it is produced, then either eating it or carrying it away for disposal some distance from the nest. In hole-nesting species this is less of a concern and overall nest hygiene standards are therefore not so scrupulous. In the case of deeper burrows the lack of sanitation can cause a build-up of rotting detritus: the fishy stench from a kingfisher hole occupied by a brood of growing youngsters can carry for some distance.

GETTING FED

Feeding a brood of demanding and seemingly insatiable chicks requires a tremendous investment of energy on the part of the parents, who will spend all the hours of daylight searching for food and taking it back to the nest. Even after the young have fledged and left the nest they continue to be reliant on their parents, either sitting around nearby and waiting for the adults to return with food or actively following them around, demanding to be fed. Passerines

Above Young albatrosses like this **Grey-headed** *Thalassarche chrysostoma* face long periods alone, awaiting the return of their parents from feeding sorties far out at sea. **Opposite** Demanding chicks can see adult birds like this **European Blue Tit** spending 14 hours or more a day gathering food.

bring the food items back to the nest every few minutes in an endless relay, literally stuffing them into the closest available mouth or mouths. As the chicks grow, so vocalization plays an important role in the feeding process. They call almost incessantly – this is especially the case in hole-nesting species, where the visual stimulus of the open gape is less meaningful and the risk of attracting predators by calling out is less significant. Other stimuli are important, too; for example, chicks of many species are programmed to beg for food from particular shapes or colours, so even crude artificial cut-outs of gull heads elicit a response from orphaned gull chicks.

Birds such as cormorants, pelicans and gannets feed their young with partly digested food, either dropping it near them or, more often, transferring it directly down their throats, with a chick reaching up so that its own head is engulfed by the parent's mouth. Seabirds often have to fly considerable distances to gather food, so return to feed their chicks only every few hours, or even longer in some cases. The sometimes lengthy intervals are compensated for by the fact that the parents are able to deliver large amounts of food on each visit. Albatross chicks are guarded carefully by their parents for the first three weeks of their lives, but after that point – by which time a chick can regulate its own temperature and defend itself against most natural predators – a youngster may spend long periods alone, awaiting the next food drop. It will often fledge in the absence of its parents.

Some birds need to prepare the food before offering it to their young. Birds of prey must first pluck or strip prey items, then tear off small pieces of flesh to pass to their chicks. There is little or no apparent effort to share the food equally – the oldest and biggest chicks muscle their way to the lion's share, leaving their smaller siblings to wither and ultimately perish. Fratricide is not uncommon either, with hungry dominant chicks sometimes turning on and consuming their weaker brethren.

Young birds sometimes require different food from that eaten by their parents. Predominantly seed-eating sparrows feed their chicks almost exclusively on insects, especially caterpillars, which are more nutritious and provide the high-energy intake essential for rapid nestling growth. An even more specialized diet is that followed by young pigeons and doves, which exist on a special secretion known as pigeon milk. The 'milk' is produced by special cells in the crop of both parent birds and regurgitated for the chicks. High in protein and fat, it comprises the sole food for the nestlings when newly hatched, but as they grow, so the parents gradually add other items from within their crops, such as seeds and vegetable matter. By the time the chicks are ready to fledge, they are following a diet virtually identical to that of their parents, which by then have stopped producing the milk.

The young of precocial species are foraging for their own food within hours of hatching, but for the first few days still rely mainly on items brought to them by their parents. This process enables them to identify appropriate food matter for themselves and learn where to forage and what to look for. Like altricial hatchlings, they call to their parents regularly to stimulate them to bring food. Birds such as gulls and terns are regarded as semi-altricial, in the sense that their chicks hatch in a precocious state, with eyes open and covered in down, but remain in the nest and rely on their parents to feed them until they are fledged.

The young of brood parasites such as cuckoos and cowbirds lead their own paths through the processes of hatching and being reared. A young chick of the Common Cuckoo ejects its host's eggs by pushing underneath them, lifting them up one by one into the hollow of its back and tipping them out of the nest. As the only chick left in the nest, it is guaranteed all the food the host parents can provide. Other brood parasites may live alongside their host's young, but must give the right signals to their foster parents so that the host's own chicks are not fed in preference to the usurper.

Top left Raptor chicks can be very competitive and sibling rivalry is common, as shown by these young **Steppe Eagles** *Aquila nipalensis*. **Top right** **Little Tern** *Sternula albifrons* chicks hatch in a well-advanced state but rely on their parents for food. **Above** Precocial chicks like this **Demoiselle Crane** *Anthropoides virgo* learn what to eat by copying their parents. **Overleaf** **Eurasian Reed Warblers** *Acrocephalus scirpaceus* are a common host species for the **Common Cuckoo**. The baby cuckoo will have ejected the warbler's own eggs from the nest.

PROTECTING EGGS AND CHICKS

For birds nesting on the ground, safety from predators relies heavily on the camouflage provided by the incubating bird's plumage and upon the cryptic patterning of the eggs or chicks. Adult birds are constantly vigilant and, when they perceive danger approaching, give an alarm call. In the case of waders, gulls and terns this is a signal for the chicks to lie flat and remain completely motionless until given the 'all clear'. Protective parents also employ specific strategies to confound would-be predators. Foremost among these is the so-called distraction display of plover species, in which an adult bird feigns injury by dragging its wing as if broken and scuttling along the ground, thereby luring the predator away from the eggs or chicks. Once the predator is safely removed from their vicinity, the adult bird leaps up and flies off.

Some birds attempt to relocate their young when danger is present. This is not an option for most species, but birds of prey are known to pick up their chicks in their feet and move them from one place to another, whilst rails and gallinules have been recorded doing so by picking up the nestlings in their beaks. One of the strangest potential means of transportation is by woodcocks *Scolopax* spp., which allegedly can carry their young in flight by cradling them against their underparts using their legs. There are many anecdotal reports of such behaviour, but it remains to be proven scientifically.

When their eggs or young are faced with danger, some species go for direct confrontation. Arctic Terns *Sterna paradisaea*, skuas or jaegers (family Stercorariidae) and owls may all swoop down on potential predators near their nests and actually make physical contact using their wings, bills or feet. They have even been known to draw blood from the heads of people walking close to their eggs or chicks. Similar physical action is taken by a range of birds against predators like corvids, and parasites such as cuckoos and cowbirds, which are 'mobbed' remorselessly whenever they are spotted near a nest. Birds will often act in concert to drive away such intruders.

The eggs of even the most attentive of bird parents still sometimes suffer misfortune. Eggs may roll out of a nest, for example, a problem to which responses vary. Passerine birds, most of which build cup nests in trees or shrubs, show no interest in retrieving an egg lost in this way. Ground-nesting ducks and gulls will, however, attempt to bring an errant egg back into their nest by reaching out, cupping their bill underneath the egg and trying to roll it back towards them.

Opposite Arctic Terns may launch physical attacks on humans that approach their eggs or young too closely.

COOPERATIVE BREEDING

The level of attention and feeding required by altricial young is such that both parents must play a part in rearing them. In the event of one parent perishing, it is almost impossible for the surviving mate to rear a brood unaided, although this has been recorded in cases where the young were already fairly advanced and close to fledging. In some cases – for example in hornbills (see page 110) – a very high level of close cooperation is required between the two parents.

In the case of some precocial species it is only ever the mother (or, more rarely, the father) that rears the young, the other parent playing no further role in the reproductive process after the act of mating. Yet at the other end of the spectrum is the scenario of cooperative or collaborative breeding, in which responsibility for the rearing of young is shared by more than two birds. Such behaviour has been noted in over 300 different species, and in most cases takes the form of 'helpers' drawn from previous broods (or siblings, in some cases), which help their parents rear subsequent generations of chicks. In birds such as the Florida Scrub Jay *Aphelocoma coerulescens* this may be an evolutionary response to a situation where there is a lack of suitable opportunities for range expansion. Florida Scrub Jays are confined to a relict population living at maximum density within the limited area of suitable habitat. With scope for establishing new territories so constrained, young jays often stay with their parents for several years, helping with feeding chicks and group-defending the territory against predators and other jay families.

In the case of the enigmatic Hoatzin *Opisthomocus hoazin*, several adults often serve as auxiliaries to a breeding pair. They help with building the nest, feeding young and watching out for predators. In this species the habitat limitation concerns of the Florida Scrub Jay do not apply, so the prevalence of cooperative breeding among Hoatzins is probably part of a complex social organization as yet not totally understood. There are clear practical benefits, however, as in the case of other cooperative breeders such as bee-eaters, where the presence of a 'helper' with feeding appears to boost the number of chicks that fledge successfully.

Anis *Crotophaga* spp., members of the cuckoo family, even breed communally. Up to four pairs live together, the females laying their eggs in one nest and sharing incubation and the feeding of hatchlings. In two of the three ani species the pair bond has broken down completely and males and females mate promiscuously with one another. Such free love has a downside, though. A dominant female usually emerges and ejects the eggs of other females in the group from the nest before settling down to lay last and brood her own clutch.

Cooperative behaviour with the protection of chicks is found in birds such as penguins, flamingos, and ducks such as the Common Eider and Common Shelduck. A few adult 'guardians' are responsible for a crèche with large numbers of young, to the vast majority of which the guardians are not related. The crèches operate on the principle of safety in numbers and allow individual parents to go off to feed or, in the case of the Common Shelduck, abandon their young to the permanent care of others whilst they migrate away to moult. Some birds appear to operate a crèche shift system, with the guardians relieving each other from time to time. In other species the guardians may be semi-permanent failed breeders or simply 'aunties' without a mate or young of their own.

Top Young flamingos (here **Lesser Flamingos**) are often gathered in crèches, looked after by 'aunties'. **Above** **Hoatzins** are almost always seen in groups, with several birds often assisting a breeding pair.

THE FINAL ACT OF INDEPENDENCE

From hatching to successful fledging and independent life can take as little as four weeks among passerines. Indeed, many such birds are multiple-brooded, so need to forge ahead quickly with their next brood once the preceding one is off their hands: the parents may simply stop feeding the fledged youngsters, thereby forcing them to look after themselves. Precocial young such as those of waders and ducks are usually also self-sufficient in a month or so. They may well remain in the general vicinity of their parents, but to all intents and purposes are independent by this time.

In larger birds the process is considerably more protracted. Their chicks take much longer to develop, so remain in a dependent state for several months in some cases. Albatross chicks may take up to a remarkable 280 days to fledge, much of which time is spent waiting for their parents to come back with food for them. They often fledge when their parents are away, leaving them to return to an empty nest.

Large birds of prey such as Golden Eagles *Aquila chrysaetos* take up to four or five years to become fully adult and sexually mature. Whilst initially the juveniles may well stay within their parents' territory, they are usually chased away by the older birds by the time the next breeding season comes around and are then forced to find a territory of their own. This marks the start of a new round in the reproductive process, with the newly independent offspring now seeking to establish itself in a tract of suitable habitat and to begin the search for a mate with which it can breed.

This **Golden Eagle** has large areas of white on its wings and tail, which mark it out as a juvenile bird. It will not be fully adult for another two years.

Glossary

allopreening Mutual preening between birds, usually (but not exclusively) a mated pair. A central element of pair bonding.

altricial Helpless upon hatching from the egg, with eyes closed and usually naked. Totally dependent upon the parents. See also **nidicolous**.

auntie Adult bird that assists with looking after young birds to which it may not be related, usually by tending them in a crèche.

billing Mutual touching and crossing of bills, often undertaken by birds such as gannets and storks as part of courtship and pair bonding.

booming Extraordinary deep and resonant sound, produced by bitterns and certain grouse species as part of their territorial/courtship display.

bower Construction, often resembling a stage, built solely for display purposes by bowerbirds. Often decorated with a wide range of objects.

brood Group of young birds from a single breeding attempt, also known as a clutch. Also used as a verb, meaning to incubate eggs or, more usually, keep young chicks warm by settling over them.

brood parasite Bird that lays its eggs in the nests of other species and plays no role in the rearing of its own young. Cuckoos and cowbirds are among the more notorious followers of this breeding system.

brood patch Area of exposed skin on a bird's abdomen that it uses to incubate its eggs and brood its young when they are first hatched. The patch disappears later in the breeding season.

cloaca Joint opening of a bird's reproductive and digestive systems and the key element in avian copulation.

clutch Group of eggs or young from a single breeding attempt. See also **brood**.

colony Group of birds that assembles in one location for the purpose of breeding. Seabirds are typical, often gathering in vast mixed colonies within which each species usually maintains its own niche.

cooperative breeding Breeding system in which the parents are assisted by other birds, usually their offspring from previous broods, in the rearing of young.

courtship Behaviour between the sexes that forms part of the prelude to copulation. Actual courtship displays are usually performed by the male.

crèche Group of young birds, often unrelated, and gathered together under the protection of a small number of adults. See also **auntie**.

cryptic In the cases of eggs, chicks or plumage: patterned or coloured in such a way as to provide camouflage against a particular background. Also refers to behaviour that is designed to confuse or evade predators.

dawn chorus Singing of large numbers of birds in the hours immediately before and after sunrise. Particularly prominent at the beginning of the breeding season.

distraction display Behaviour undertaken by parent birds, especially of wader species, when attempting to divert a predator away from their eggs or young. Usually involves feigning injury, especially by dragging a supposedly broken wing.

drumming Sound made by woodpeckers when they repeatedly strike their bills against wood as part of their territorial display. Also used for the sound produced by other bird types, including snipe and Ruffed Grouse, making similarly resonant effects via their feathers or other body parts.

egg tooth The small protruberance on the end of a hatching chick's upper mandible, which it uses to break through the shell. The egg tooth is lost soon thereafter.

fledging Acquisition by a chick of its first full set of feathers, which enable nidicolous birds to leave the nest.

fledgling Young bird that has fledged and left the nest.

hatchling Young bird that has recently emerged from its egg.

incubation Sitting on a clutch of eggs and facilitating their development via the provision of body heat; in a few species (e.g. megapodes), incubation is effected by other means, however.

lek Area of (usually) open ground on which male birds of a particular species gather to display and attract females to mate.

monogamy Breeding system in which male and female birds form a pair and mate (mostly) only with each other. They remain closely bonded, either within a particular breeding season or for life.

nestling Young bird that is still confined to its nest.

nidicolous Remaining in the nest after hatching. See also **altricial**.

nidifugous Leaving the nest shortly after hatching. See also **precocial**.

oology Study or collection of eggs.

passerine Member of the Passeriformes, comprising over half of all bird species and characterized by their foot arrangement, with three forwards-pointing toes and one backwards pointing.

penguin dance Courtship display by grebes, notably Great Crested, in which male and female rise up out of the water, facing each other, and present pieces of vegetation whilst shaking their heads. Part of courtship display and pair-bonding.

polyandry/polyandrous Mating system in which a female is mated with two or more males.

polygamy/polygamous Mating system in which a bird of one gender is mated with two or more birds of the opposite gender.

polygynandry/polygynous Mating system in which two or more males mate with two or more females.

polygyny/polygynous Mating system in which a male is mated with two or more females.

precocial Well advanced upon hatching, usually with the eyes open, covered in down and able to walk and feed shortly after emerging from the egg. See also **nidifugous**.

raptor General term for a diurnal bird of prey.

roding Performing of a territorial display flight by species of woodcock.

role reversal When typical sexual rules in breeding are reversed, e.g. with the female initiating display and the male carrying out most of the incubation, rearing the young, etc.

semi-precocial Well advanced upon hatching, but remaining in the nest and dependent on parents for food. See also **precocial**.

sexual dimorphism Established differences in size, appearance and behaviour between male and female, in which the male is usually larger, more brightly coloured and more dominant than the female. Reverse sexual dimorphism is when the opposite situation applies, i.e. female brighter, etc.

sky-dancing Raptor display flight in which the male (and sometimes the female) makes a series of marked ascents and descents, interspersed with glides and periods of hovering.

sky-pointing Part of courtship display and pair-bonding undertaken by seabirds such as albatrosses and gannets, in which both male and female face each other and point their bills vertically towards the sky.

territory Area of habitat defended by a bird or birds for the purposes of breeding, feeding or roosting.

wader General term for long-legged coastal and wetland birds such as plovers, sandpipers and shanks.

Resources

PUBLICATIONS

Few books currently in print are devoted to bird reproduction and breeding habits. The following two DVD–ROM publications are of interest.

Breeding Birds of Britain and Ireland: Nests, Eggs, Nestlings, Fledglings and Habitats, Peter Castell & Richard Castell, BirdGuides (2009).

Breeding Birds of the Western Palearctic: Nests, Eggs, Nestlings, Fledglings and Habitats, Peter Castell & Richard Castell, BirdGuides (2009).

BIRD ORGANIZATIONS

With public interest in ornithology at a record high in many countries, the major bird conservation organizations now offer opportunities to learn more about avian reproduction. These range from organized trips to view birds on their breeding territory, without the risk of disturbing them, to the chance of undertaking survey work on various breeding species. See below for further details, as well as general information about different bird species, their nesting habits and the code of behaviour to follow when near birds, their nests and young.

The Royal Society for the Protection of Birds
www.rspb.org.uk

The British Trust for Ornithology
www.bto.org

The National Audubon Society
www.audubon.org

Index

Page numbers in **bold** indicate illustrations. Within the book the scientific name is generally only provided on first mention of a species.